T0228108

LIVING WITH EPIDEMICS IN
COLONIAL BENGAL

LIVING WITH EPIDEMICS IN COLONIAL BENGAL
1818–1945

ARABINDA SAMANTA

Routledge
Taylor & Francis Group

LONDON AND NEW YORK

First published 2018
by Routledge
4 Park Square, Milton Park, Abingdon, Oxon OX14 4RN
605 Third Avenue, New York, NY 10017

First issued in paperback 2023

Routledge is an imprint of the Taylor & Francis Group, an informa business

Publisher's Note
The publisher has gone to great lengths to ensure the quality of
this reprint but points out that some imperfections in the
original copies may be apparent.

Print edition not for sale in South Asia (India, Sri Lanka, Nepal, Bangladesh,
Afghanistan, Pakistan or Bhutan).

British Library Cataloguing in Publication Data
A catalogue record for this book is available from the British Library

Library of Congress Cataloging in Publication Data
A catalog record for this book has been requested

ISBN-13: 978-1-138-09535-9 (hbk)
ISBN-13: 978-1-03-265255-9 (pbk)
ISBN-13: 978-0-203-73138-3 (ebk)

DOI: 10.4324/9780203731383

Typeset in Adobe Garamond Pro 11/13
by Ravi Shanker Delhi 110 095

MANOHAR

To

PROFESSOR DEEPAK KUMAR

Contents

Preface

This volume is a revised and updated version of my post-doctoral work on popular perceptions of disease and medicine. It engages with multiple epidemics that gripped the Indian subcontinent through the nineteenth and early twentieth centuries. The spatial location is, however, colonial Bengal, though for obvious reason the focus shifts to other region of this subcontinent as well. The project was funded by the Wellcome Trust for the History of Medicine, University College, London, as also by the Zakir Husain Centre for Educational Studies, Jawaharlal Nehru University, New Delhi. A research grant from Rockefeller Foundation enabled me to consult documents housed in the Rockefeller Archive Center, New York.

Much of what I argue in this book has been presented at various national and international seminars/conferences during the last decade, and it is only after adequately addressing my fellow colleagues' informed criticism, and sustained by their constant prodding, that I ventured to put them in this present form.

The major arguments in the chapter on Plague were presented at the International Conference on 'Hybrids and Partnerships: Comparing the Histories of Indigenous Medicine in Southern Africa and South Asia', organized jointly by Manchester University and the Wellcome Trust for the History of Medicine at Oxford, held at Oxford on 15-16 September 2005. The paper drew interested attention of Professor Mark Harrison, and I am extremely grateful to him for his valuable comments on the earlier draft of the paper. Part of this presentation was published in Ranjan Chakrabarti, ed., *Situating Environmental History* (2007) and a different version of it was included in Arun Bandopadhyay, ed., *Science and Society in India, 1750-2000* (2010).

The present chapter is, however, thoroughly revised and updated, and argued on a much comprehensive understanding of the theme.

The chapter on malaria draws heavily on my Doctoral dissertation, but has been substantially expanded and structurally reformulated. It was presented to the 18th IAHA Conference, held at Academia Sinica, Taipei, Taiwan, 6-10 December 2004. I thank Professor Kohei Wakimura, Osaka City University and the Discussant of the session, who rated my paper on malaria as 'a pioneering work'. A different version of this piece has been published in Poonam Bala, ed., *Contesting Colonial Authority: Medicine and Indigenous Responses in Nineteenth- and Twentieth-century India* (2012).

My work on smallpox was originally presented in lesser form to a seminar on 'Public Health in India: Colonial and Post-colonial Experience', organized jointly by Institute of Development Studies, Kolkata and Department of History, Calcutta University on 15-16 October 2003. I thank Professor Bhaskar Chakrabarti, my friend in the Department of History, Calcutta University, who suggested a few important changes. More nuanced and elaborated, it was presented to the European Social Science and History Conference, held at Lisbon, Portugal in February 2008. A different version of this paper was published in the *Indian Journal of History of Science* in its July 2011 issue.

The present study is based on various source materials housed at Oriental and India Office Collections, British Library, London, the Wellcome Trust for the History of Medicine Library, London, Rockefeller Archive Center, New York, Indian National Archives, New Delhi, West Bengal State Archives, Kolkata, National Library, Kolkata, School of Tropical Medicine Library, Kolkata, Centre for Studies in Social Sciences Library, Kolkata, Bangiya Sahitya Parishad Library, Kolkata, Jaykrishna Library, Uttarpara. I owe special thanks to the staff members of all these repositories of information who gave me ready access to various sources.

I shall be failing in my duty should I not record my heart-felt gratitude to quite a few other special individuals who helped me in my work in many ways. I am deeply indebted to Professor Sekhar Bandyopadhyay, my first PhD supervisor at Calcutta University, who introduced me to historical research; I am also immensely grateful to late Professor Basudeb Chattopadhyay under whom I concluded my dissertation when Professor Bandyopadhyay left for Victoria University at Wellington. Both of them helped me conceptualize a research project and contributed significantly to my understanding of history. My friends in Kolkata, Professor Suranjan Das, Professor Smritikumar Sarkar, Professor Arun Bandopadhyay, Professor Ranjan Chakrabarti, Professor Sujata Mukherjee, and Professor Bipasha Raha helped me through discussions to enrich my understanding and sharpen my arguments. Bhaswati, my wife, remained all along extremely supportive as ever. I owe much to her for her lost weekends. She had volunteered to live with me for months together in America with our daughter Samiparna to give me respite from household chores and focus on this book. Tannistha and Samiparna, my twin daughters, never tired of enquiring about the progress of the work even though both had been extremely hard-pressed with their academic work. Samidh, my son-in-law and Atlanta-based data scientist gave me as usual a welcome relief from my 'disease hunting' in archives with innocuous jokes via e-mail. Kaushik, son-in-law and ISRO scientist, maintained an inspirational role with his saintly reticence and occasional verbosity. Achintya, Sudit, Malabika, Tanveer, Aparajita, Pradip, Binata, and Suvo, all colleagues in the Department of History, Burdwan University, Anuradha, Sudeshna and Maroona in the Department of History, Jadavpur University, Gautam and Debjani at Vidyasagar University and Subrata at IDSK were friendly and cooperative, bearing many of my foibles. I should also express my deep regard for Mrs. and late Mr. Pundit who made my stay in London comfortable. I am also grateful to my anonymous referee, who raised some pertinent questions on an earlier draft of this work.

And last but not the least, Professor Deepak Kumar, Professor of History of Science and Education and Concurrent Professor, Centre of Media Studies, School of Social Sciences at Jawaharlal Nehru University, New Delhi, who initiated me in the study of epidemics and medicine, remains my perennial source of inspiration as ever. It was no less a person than Deepakda who mooted the idea of working on this project, provoked me and offered me invaluable support. The present book owes much to his patient hearing and intellectual input. I acknowledge my debts of gratitude to him. The errors, if any, are entirely mine. I would also thank Mr. Ramesh Jain of Manohar for his indulgence in granting me enormous time for preparing the manuscript and finally publishing this work.

ARABINDA SAMANTA

Introduction

Not infrequently, disease has informed the trajectory of human history over the world. Epidemics have always struck terror in society and have left dramatic impact. Human society has always been overwhelmed by the devastating blow of bacteriological invasions all the way from Periclean Athens through Black Death in the fourteenth century in Europe to sub-Saharan Africa in the twentieth century. Diseases have indeed had such a compelling effect on the contemporary society that Western historiography in the past three decades has brought to bear the notion of 'social construction' of disease in its accounts and sought to explicate how people of the West understood, coped with, and tried to prevent various infectious diseases in their several ways.

In contrast to the range of scholarship now available on the social history of disease in Africa, Europe and North America, for South Asian history, disease remains a relatively less addressed field. India proved to be the largest disease laboratory in Asia throughout the entire colonial period. Malaria, smallpox, plague, and cholera were regular visitors to the subcontinent, sometimes with benign intimidations, but more often than not with epidemic vengeance. Thousands of people used to vanish within days, leaving behind those who managed to survive all but maimed in body and shattered in soul. Though monographs and scholarly papers on such diseases are coming up in abundance in recent years, the biography of a particular disease with its multi-spatial locations still remains an important desideratum.

Any historical engagement with epidemic diseases calls for an attempt to situate their locations. A disease is not just an event involving only microbe/virus/bacteria and its host, a patient and a doctor, a medicine and its remedial property. It also strikes in different places and different times.

Arguably, three major issues can be identified. The first is the nature of intervention by a government and its acts of omission and commission to address an epidemic disease. The way it responded, the resources it mobilized to arrest its progress, and the method it employed to combat the disease, all go to inform this location. The second issue is the illness itself. How did the disease spread? Did it 'invade' too many people within too short a time? How did it progress? Did it prefer a particular geographical location or a particular class or community? A disease provides a space for interaction between a patient and a physician. How does a patient evaluate a doctor? Likewise, how does a doctor engage his patient? And finally, the most crucial issue is the whole host of people affected by the steady and progressive expansion of an illness at a certain time.

In fact, the most important question is how did the people perceive those epidemic diseases? Was their behaviour based on more simplistic vector/host concepts that are so common in infectious disease fields? Or were the diseases perceived as metaphors?[1] Generally missing in recent discourses on epidemic and medicine is the patients' perception. A patient, for instance, may complain of symptoms such as nervousness, fatigue, loss of appetite, sleeplessness, or vomiting. To a practitioner of 'modern' medicine, this 'vague' symptomatology might amount to a different semiotic category: he might correct the patient and suggest that he is suffering from 'tubercular' disease. Thus the doctor's diagnosis posits an anatomical transposition: from nerve to lung. The subjective perception of the patient is thus re-formulated into an objective description by the physician.[2] To a physician, thus, the patients remain no longer individuals with their own notions of disease symptomatology; they increasingly become 'cases'. The disease itself has become more important than the sick person.[3]

The moot question is, therefore, how people in the past perceived a killer disease. Some scholars have looked into magico-religious rites and rituals in connection with epidemics in colonial India.[4] But the problem is how these can be contextualized historically, how one can trace the interface between the great tradition and little tradition in terms of folk medicine and ethno-history.[5] Larger than this is the more crucial question how did the sick evaluate a doctor? Recent historiography maintains a rather studied reticence on such questions.

Although we have had a few monographs on some epidemics – malaria, cholera, plague and small pox, for instance, we are yet to know adequately the popular perception of these diseases. Nevertheless, disease historiography has witnessed multiple shifts in recent years. Initially, attention was paid to medical policy and medical policing. Scholars sought to show how the imperialist powers administered matters of health and medicine as 'tools of empire',[6] to reduce mortality in their tropical colonies. Mortality rates for Europeans, it was argued, were much higher in colonies like India than in Britain.[7]

But the rhetoric of medical intervention was selective and piece-meal. It addressed the health of the troops and the European civilians, and as an after thought health of the indigenous people.[8] It has been argued that the most important cause of sickness among the soldiers was sexually transmitted disease. Unable to discourage their men from visiting prostitutes even through religious persuasion, the authorities targeted the brothels.[9] The colonial government of India introduced in 1868 the Contagious Diseases Act, which largely informed the racial perception of diseases.[10] This legislative intervention was construed by Indians as infringing on their religious sensibilities and civil liberties.

This brings in a fresh debate among historians concerning medical intervention and the response it generated in Indian society. Recent historiography has tended to shift focus to attempt to explore other dimensions of 'colonial medicine'. Mark Harrison has rightly questioned 'what was specifically colonial about "colonial medicine"? 'Should we speak of "medicine in the colonies", rather than of "colonial" or "imperial" medicine?' Particularly important in this respect is the relation with colonial power/knowledge, as also the response of local communities according to class, caste, ethnicity and gender.[11] In India it was particularly so, and Mark Harrison does admit it, because, he argues, 'decision-making' in matters of medical and sanitary policy increasingly came from within the competence of Indians themselves. But what he seems to have missed is that this was a much later development, say after the First World War. During the late nineteenth century, as local municipal bodies were coming up to share power/knowledge with the ruling authority, there was proliferation of major epidemics causing havoc with the indigenous population. Crucial decisions regarding the nature of prophylactic intervention to combat the epidemics and their method of execution rested with the imperial authority alone. The epidemic plague in Bombay is a case in point.[12]

More recently, historians of medicine and public health have brought out three main historical questions: epidemics as causative agents of change, epidemics as mirrors reflecting social processes, and epidemics as ways of illustrating changing medical theories and practices.[13] The present work tries to combine all three of these approaches, but at the same time adds to it a fourth element. How did the people react to such epidemics? In other words, it seeks to explore the interface between racism, imperialism, capitalism, poverty and disease.

There have been in more recent times some serious attempts to engage historically some of the killer epidemics in certain parts of India.[14] Simkie Sarkar, for instance, has engaged with malaria in nineteenth-century Bombay.[15] She is basically concerned with the disease as such, the agent of its transmission, the way the government sought to prevent it, and the economic cost analysis of the disease. Ihtesham Kazi sought to locate malaria in the environmental decay of colonial Bengal.[16] V.R. Muraleedharan recounts the story of malaria in Madras and the colonial government's response to it in the early twentieth century.[17] Dhrub Kumar Singh has tried to dispel 'clouds of cholera' and clouds around cholera in India.[18] Within the climate paradigm and under the influence of a theory of multiple causes, cholera, Singh argues, introduced in colonial India an element of uncertainty and fluidity and also shifting standpoints in medical discourse. In all these scholarly engagements, colonial governments seem to persist in the centre stage of their narratives. People, it appears, remain peripheral players.

Another round of research initiated by Mark Harrison and Biswamoy Pati seems to inform a fresh look into disease disorder.[19] The contributors to this work attempt to demonstrate the complexity of relationships between colonizers and colonized, and the diversity of colonial 'impacts' on indigenous society. Sanjoy Bhattacharya, for instance, questions the habit of locating Western medical intervention as a 'monolithic' medical establishment.[20] In his stimulating piece on smallpox vaccination, he argues that state medical interventions in India have always had very different histories in the urban and rural contexts, a phenomenon not adequately addressed in the extant historiography. Bhattacharya emphasizes the fact that while important differences persisted in separate urban contexts to official medical interventions, the vaccinating establishments could better target the towns and cities, especially because they had the advantage of operating in a limited geopolitical boundary. By contrast, the countryside presented difficulties of scale, and the local vaccinating establishments had to struggle hard to seek better directions.[21] He contextualizes the government vaccination programme by referring to rural Indian social factors as also infrastructural constraints which tried to explain much of its non-performance. But what of the perception of the people for whom the entire endeavour of the government was geared?

Biswamoy Pati and Manjiri N. Kamat have examined the multiple facets of colonial interventions at two pilgrimage sites, Puri in Orissa and Pandharpur in Maharashtra, occasioned by the concerns of cholera in the former and plague in the latter.[22] Government initiative in ameliorating the cholera situation in the 'valley of death', argues Pati, was weak, indecisive, contradictory and lacked serious direction. This was presumably because the 'valley of death' rarely posed a threat to the colonial army in the early nineteenth century. That Puri as a town was left unplanned and in a state of 'disorder' related to the process of its colonization, and although colonial health interventions targeted the pilgrim routes, the obsession with Puri and the *ratha yatra* prevented any possibility of actually going out of Puri town and intervening seriously in the agenda of health in the district. Pati appears to locate his account in a synthesized perception of three-tiered sites, the missionaries, the colonial administrators, and the so-called enlightened 'native'. The unlettered 'native' does not seem to fit in his scheme of interrogation. Manjiri Kamat seeks to explain the connection between two fairs at

Pandharpur against the backdrop of plague and cholera epidemics.[23] She is primarily concerned with the official position which she calls 'strategies of evasion'. She seeks to show how differences in threat perception, colonial imperatives, and religious prerequisites went into colonial understanding of the 'order', and how the local interest groups including the 'powerful priestly families', local officials, and the Bombay government were enmeshed in a contradictory relation in all sorts of medical intervention. Indigenous popular response to colonial interventionist endeavours is downplayed in interrogation.

Still more recently, public health, race, gender, medicine, Indian medical traditions and many more issues inform Pati and Harrison's social history of health and medicine in colonial India.[24] Sanchari Dutta engages plague, quarantine and empire in the context of British-Indian sanitary strategies in Central Asia, while Projit Bihari Mukharji revisits 'indigenous knowledge' as an 'epitaph of subaltern science'. Achintya Kumar Dutta explores the nature of medical research on kala-azar and its control in colonial Assam, while Biswamoy Pati and Chandi P. Nanda attempt a social history of leprosy patients in colonial Orissa. The contributions offer valuable insight into diverse themes but none seems to add to our knowledge about Bengal proper with regard to the diseases with which the present study is concerned.

As against the notion of 'colonizing the body', sustained for several decades by scholars of medical history, one notices now a remarkable positional shift to 'nationalizing the body', which refuses to see 'Western' medicine as an alienated adjunct of the colonial state.[25] Projit Bihari Mukharji argues that a progressive medical market, and a medical publishing industry, gave *daktari* medicine a distinct social identity not entirely associated with the state. Mukharji seeks to understand how *daktari* medicine, as distinctly different from 'Western medicine', tended to reposition the colonized bodies as nationalized bodies.

In interaction with the colonial authority local responses 'produced new forms of structures', says Poonam Bala,[26] which she calls 'paradigms of defence'. True, contestation, acceptance, and resistance were obvious outcomes of such encounters, but truer still, diseases involved attributes other than the medical and invested disease with different social meanings. To combat epidemics, much involvement and compliance of local population was necessary. One therefore needs to locate diseases primarily at the popular level. In a more recently published book,[27] Bala seems to be aware of this problem; she admits that 'explaining public health issues, colonial attitudes and Indian resistance ... has been the pivot of analysis for medicine and its engagement in colonial India.' Instead, she seeks to showcase 'the common theme of intellectual engagements of medicine with indigenous populations and their already existing institutional paradigms'. But the problematic persists.

The truth in the historical process of colonization and its inscrutable theme of disease importation, Deepak Kumar argues rightly, is located somewhere between 'the changing patterns of the diseases, their impact on society, and the socio-cultural ways in which people relate to health issues'.[28] The interaction between the social and medical domains was seldom linear, very often it posited conflict, difference, and at times aggression. In all such eventualities, the victim, the patient should surely be the key to the historical narrative.

II

The moot question is, however, how to access historically the perception of the people. To put it differently, what would be our tool of analysis? A variety of perspectives has been employed in research on popular perceptions of risk.[29] While the 'knowledge theory', for instance, argues that people perceive things to be dangerous because they 'know' them to be dangerous, the 'personality theory' suggests that individuals are so constituted as to take or avoid risks in acceptance or passivity rather than courage or patience. A third approach follows 'the economic theory' which contends that the rich are more willing to take risks because they are somehow shielded from their adverse consequences, and that the poor presumably feel just the opposite, because they are more vulnerable. 'The political theory' views risks as struggles over interests, such as holding office or party advantage.

The present study is not primarily addressed to any of these theories.

It draws heavily on social, cultural and linguistic anthropology to understand the functions of health experiences, distribution of illness, prevention of sickness, social relations of therapeutic intervention and employment of pluralistic medical systems. It interrogates the social construction of medical knowledge, politics of science, and the changing paradigm of relationship between health of the individual and the prerogatives of larger colonial economic formations.

Nevertheless, I view individuals as the active organizers of their own perceptions. Basically a historical interrogation, the position is of a cultural theorist proposing that individuals choose what to fear and how much to fear in order to support their own way of life.[30] While I acknowledge that personality, political orientation or demographic variables are all important, I find that cultural theory provides the best prediction of a broad range of perceived risks and an interpretative framework in which the following findings cohere.

The four major chapters in this present study are about the social and political construction of four crucial diseases, viewed in the perspectives of political economy of the state and the cultural fixities of the victims, holding the people hostage through the nineteenth and twentieth centuries. Perhaps selective engagement with only four diseases calls for an explanation. Smallpox, plague, cholera, and malaria, one can argue, were the major killers that held the people in perpetual fear. People were afflicted by other killer diseases as well, but the victims seemed to have recovered without impaired physical functions. It was the diseases in question that visited Bengal all on a sudden. They caught people unawares, killed them in thousands, and changed a society and its demographic structure substantially. This is why these major killers compel our attention.

The study thus puts diseases in a social, cultural and political context. It seeks to explore how, sometimes through mutual adaptation but more often by cultural contestation, people pulled on with others similarly stricken. To Indians, diseases were often constructed by certain myths and metaphors, considered by many as curse of gods or goddesses or divine punishment, certainly an embarrassment to their neighbours in the rural society. In most cases, illness became a metaphor for the dangers of an improper code of conduct that needed to be corrected by a personal expropriation of the sin committed, or by community worship of the deity supposedly responsible for it. As a result, 'Western' medical science often took a back seat, and elaborate rites and rituals with supposedly curative values were observed. Epidemics were also interpreted as outcomes of politically incorrect moves made by a ruling power. To right the wrongs, people very often resorted to social protest. The protests by the Indian literati were sometimes muted, but turned violent when those who had no axe to grind took up the cudgels.

Although, the whole of colonial India is admittedly the scenario of this historical engagement, Bengal is the focus because it was here that two major epidemics, malaria and cholera, came in the 1820s. Furthermore, Bengal was the seat of colonial administration where official perceptions about the epidemic diseases seemed to have informed the government's attitude and helped devise a policy to combat the epidemics for the first time. The year 1818 is a convenient entry point of this study because it was around this year that British colonial administration consolidated its position through the defeat of the virtually last political contestant, the Marathas. Moreover, it was around this point of time that the two major diseases, malaria and cholera broke out as dreadful epidemics. Initially, the colonial government sought to contain the epidemics through palliatives rather than prevention. There was, however, a marked change in government policies during World War II. As soldiers began to be afflicted and consequently the course of the War was affected, one notices a remarkable shift in the colonial health policy.[31] A new sense of urgency seems to have replaced the indifference of the past. So the study ends in 1945, the end of the Second World War.

NOTES

1. Rohan Deb Roy, ' "An Unseen Awful Visitant": The Making of Burdwan Fever', *Economic and Political Weekly*, vol. 43, nos. 12 and 13, 22 March-4 April 2008.
2. Hugh Shapiro, 'Hybridic Languages of Bodily Distress: Diagnosis and Cultural Practice in Early Twentieth-Century China'. Paper Presented to the ASHM Conference on Health and Medicine in History: East-West Exchange, JNU, New Delhi, 2-4 November 2006.
3. K. Figlio, 'The Historiography of Scientific Medicine: An Invitation to

Human Sciences', *Comparative Studies in Society and History*, vol. 19, 1977, p. 265.

4. Ralph W. Nicholas, 'The Goddess Sitala and Epidemic Smallpox in Bengal', *Journal of Asian Studies*, vol. XLI, 1981, pp. 21-44.

5. Deepak Kumar, 'Social History of Medicine: Some Issues and Concerns', in Deepak Kumar, ed., *Disease and Medicine in India*, New Delhi: Tulika, 2001, pp. xiv-xvii.

6. D.R. Headrick, *The Tools of Empire: Technology and European Imperialism*, Oxford University Press, 1981.

7. Philip D. Curtin, *Death by Migration: Europe's Encounter with the Tropical World in the Nineteenth Century,* Cambridge: Cambridge University Press, 1989; See also his *Disease and Empire: The Health of European Troops in the Conquest of Africa*, Cambridge: Cambridge University Press, 1998.

8. Kabita Ray, *History of Public Health: Colonial Bengal 1921-1947*, Calcutta, K.P. Bagchi & Co., 1998; Anil Kumar, *Medicine and the Raj: British Medical Policy 1835-1941*, New Delhi: Sage, 1998. Deepak Kumar, *Science and the Raj*, New Delhi: Oxford University Press, 2000.

9. Douglas M. Peers, 'Soldiers, Surgeons and the Campaigns to Combat Sexually Transmitted Diseases in Colonial India, 1805-1860', *Medical History*, no. 42 (2), April 1998, pp. 137-60..

10. David Arnold, *Colonizing the Body: State Medicine and Epidemic Disease in Nineteenth Century India*, Berkeley, Los Angeles, London: University of California Press, 1993.

11. Ibid. Mark Harrison, *Public Health in British India: Anglo Indian Preventive Medicine, 1859-1914*, Cambridge: Cambridge University Press, 1994. Poonam Bala, *Imperialism and Medicine in Bengal: A Socio-Historical Perspective*, New Delhi and London: Sage, 1991.

12. Arabinda Samanta, 'Plague and Prophylactics: Ecological Construction of an Epidemic in Colonial Eastern India', in Ranjan Chakrabarty, ed., *Situating Environmental History*, New Delhi: Manohar, 2007.

13. Nancy Gallagher, *Medicine and Power in Tunisia, 1780-1900,* New York: Cambridge University Press, 1983.

14. Deepak Kumar, ed., *Disease and Medicine in India: A Historical Overview,* New Delhi: Tulika, 2001.

15. Simkie Sarkar, 'Malaria in Nineteenth-Century Bombay', in Deepak Kumar, ed., *Disease and Medicine in India: A Historical Overview,* New Delhi: Tulika, 2001, pp. 132-43.

16. Ihtesham Kazi, 'Environmental Factors Contributing to Malaria in Colonial Bengal', in Deepak Kumar, ed., *Disease and Medicine in India: A Historical Overview,* New Delhi: Tulika, 2001, pp. 123-31.

17. V.R. Muraleedharan, 'Malady in Madras: The Colonial Government's Response to Malaria in the Early Twentieth Century', in Deepak Kumar,

ed., *Science and Empire: Essays in Indian Context*, Delhi: Anamika, 1991, pp. 101-12

18. Dhrub Kumar Singh, '"Clouds of Cholera" and Clouds around Cholera, 1817-70', in Deepak Kumar, ed., *Disease and Medicine in India: A Historical Overview*, New Delhi: Tulika, 2001. pp. 145-65.

19. Mark Harrison and Biswamoy Pati, *Health, Medicine and Empire: Perspective on Colonial India*, New Delhi: Orient Longman, 2001.

20. Sanjoy Bhattacharya, 'Re-devising Jennerian Vaccine?: European Technologies, Indian Innovation and the Control of Smallpox in South Asia, 1850-1950' , in Mark Harrison and Biswamoy Pati, *Health, Medicine and Empire: Perspective on Colonial India*, New Delhi: Orient Longman, 2001.

21. For details see book review by Arabinda Samanta in *The Calcutta Historical Journal*, vol. 26, no. 2, July-December 2006, p. 119

22. Biswamoy Pati, '"Ordering" "Disorder" in a Holy City: Colonial Health Interventions in Puri during the Nineteenth Century', in Mark Harrison and Biswamoy Pati (eds), *Health, Medicine and Empire: Perspective on Colonial India*, New Delhi: Orient Longman, 2001, pp. 270-98.

23. Manjiri N. Kamat, 'The *Palkhi* as Plague Carrier: A Note on the Pandharpur Fair and the Sanitary Fixation of the Colonial State (1908-1916)', in Mark Harrison and Biswamoy Pati (eds), *Health, Medicine and Empire: Perspective on Colonial India*, New Delhi: Orient Longman, 2001, pp. 299-316

24. Biswamoy Pati and Mark Harrison, eds., *The Social History of Health and Medicine in Colonial India,* New York: Routledge, 2009.

25. Projit Bihari Mukharji, *Nationalizing the Body: The Medical Market, Print and Daktari Medicine in Colonial Bengal, 1860-1930*, London: Anthem Press, 2011.

26. Poonam Bala (ed.), *Contesting Colonial Authority: Medicine and Indigenous Responses in Nineteenth- and Twentieth-Century India*, Lanham, Boulder, New York, Toronto, Plymouth: Lexington Books, 2012, p. ix.

27. *Medicine and Colonialism: Historical Perspectives in India and South Africa*, ed. Poonam Bala, London: Pickering and Chatto, 2014, p. 4.

28. Deepak Kumar and Raj Sekhar Basu (eds.), *Medical Encounters in British India*, New Delhi: Oxford University Press, 2013, p. 16.

29. Aaron Wildavsky and Karl Dake, 'Theories of Risk Perception: Who Fears What and Why?', in Susan L. Cutter, ed., *Environmental Risks and Hazards*, New Jersey: Prentice Hall, 1994, pp. 166-77.

30. Mary Douglas, *Essays in the Sociology of Perception*, London: Routledge and Kegan Paul, 1982.

31. Poonam Bala, *Imperialism and Medicine in Bengal: A Socio-Historical Perspective*, New Delhi and London: Sage, 1991.

CHAPTER 1

Malaria

ANTIQUITY AND INCIDENCE

Malarial fever was in fact not unknown to the Indian people. India's acquaintance with it can be dated back to the Vedic past, for some of the earliest references to this ailment occur in the *Atharva Veda*. The *Atharva Veda* classifies the fever according to the periodicity of its attack, i.e. *sadandin* or quotidian, *tritiyakam* or tertian, and *vitrtyiyan* or quartan.[1] It also classifies the fever according to season: *sitam* or winter, *graisman* or summer, *varsikam* or monsoon.[2] It appears from the *Atharva Vedic* evidence that such fevers were common in regions like Gandhara, Anga and Magadha.[3] In all probability even the relations between mosquito and malaria were not unknown to the Vedic people, for the text gives details about mosquitoes such as the needle-like proboscis or *kusala*, the bloody mouth or *lohitasyam*, and visiting the dwellings in the dusk or *salahparinrtyantisayam*.[4] There is also reference to the use of odorous or fumigating medications to ward off the insects.[5]

Later medical commentators like Charaka and Susruta used the word *jwara* for fever in general, and henceforth all diseases with temperature as the predominant symptom are described as *jwara* by the Ayurvedic authors. The *Charaka Samhita*, written in approximately 300 BC and the *Susruta Samhita*, written in about 100 BC, refer to diseases where fever is the major symptom. The *Charaka Samhita* compares the advent and periodicity of this fever to the seed sown in the ground.[6] It takes a day to grow in case of the quotidian variety (*anyedyuskah*), two days in case of the tertian (*trtiyakh*) and three days in case of quartan (*caturthakah*). Mature seeds then invade the body and cause fever unchecked by any antibody. When the force of the invading elements is worn out, they return to their original habitat and again begin to grow. Susruta compares the advent of fever to the flood and ebb tides of the ocean.[7] Both *Charaka* and *Susruta* refer to this fever as an ailment common in lowlands or in the foothills. In all probability, Indians were familiar with the use of mosquito nets as is evidenced by the accounts of Marco Polo in the thirteenth century.

It is therefore reasonable to believe that malaria as a disease and its association with the mosquito was known since the beginning of Indian civilization, but there is hardly any evidence to prove that the disease had ever been an epidemic prior to the advent of British rule.[8] Unable to cope with the medicines at their disposal, people had learnt to live with it. But in the period under review the disease turned into an epidemic and caught the people unaware.

The predispoisng causes of malaria epidemic in Bengal emanated from the topography of the riverine areas of the province and the insanitary habits of the people. But the proximate causes stemmed largely from colonial infrastructures—roads, railroads, embankments, system of labour migration, changes in the crop pattern, export oriented commercial crops, etc. Environmental decay stemmed from long-term evolution, but British colonial policy hastened the process and accentuated the epidemic.

Environmental decay could be observed in river decay and siltation. The process was quickened by obstructed drainage which again was caused *inter alia* by railways and embankments that came in the wake of so called British development policies. Increasing number of embankments and railways reduced the chances of flood and innundation, which meant progressive loss of fertility of the soil. The consequences were generally two-fold: first, impeded drainage brought in its trail large number of stagnant pools and marshes, which provided congenial breeding ground for mosquitoes, and second, loss of fertility of the soil due to cessation of inundation diminished the amount of winter harvest, which meant starvation to many. Scarcity of food grains was further aggravated by an incentive to increased production of jute, which held out a lucrative export market after the opening of the Suez Canal. Rotting of jute plants in stagnant pools or ponds made the water still more foul and unhealthy. Mosquitoes generally preferred breeding in stagnant pools, and if such rotten pools abounded in the vicinity of a homestead, spread of malaria was ensured.

This dramatic transformation of the incidence of the disease raises a few pertinent questions. How did the people react to the epidemic? Was it with an attitude of passivity and helpless resignation? Did people blame the government, or did they try to save themselves by propitiating the gods and goddesses? These are some of the questions the present chapter seeks to raise, and by way of explanation it will address the larger question of its location in popular imagination as a disease and its remedy.

There has been a remarkable spurt in literature on malaria in recent years. David Arnold, in his work on malaria and tribal India, reminds us that colonial medical and topographical texts repeatedly identified 'fever' as a primary attribute of the environment.[9] Ira Klein revived interest in malaria and mortality in Bengal through his painstaking research on the subject.[10] Marking the Madras Presidency as his entry point, Muraleedharan has made some significant observations on the colonial government's response to changing medical theories.[11] Some recent scholars have questioned the very notion of 'malaria', arguing that in the second half of the nineteenth century in Bengal, 'a malarial epidemic' presented itself as a flexible medical metaphor: medical professionals frequently used the term to explain multiple varieties of physical ailment.[12] There have also been some attempts to explain the resurgence in term of environment.[13] More recently, scholars have also tried to invest multiple meanings with the term 'malaria' in nineteenth-century Bengal.[14] Yet, the historical trajectory of malarial fever in Bengal and its negotiation by colonial and indigenous medicine has not yet been adequately addressed. It is worthwhile to look at the functions of the disease at different locations.

POPULAR CONSTRUCTIONS

Let us identify four components among the people: the ordinary people, the rural gentry, the middle class intelligentsia, and the indigenous medical practitioners based in the villages. We shall examine the different perceptions of these four groups to the disease and how to cure it. The major emphasis will be on the perception of the ordinary people, the masses and the worst sufferers.

When the widespread epidemic malaria started taking its toll in the 1860s, popular reaction in Bengal did not differ sharply from that in other parts of the country. In fact, the identification of epidemic disease with divine wrath was almost a pan-Indian phenomenon. It took the distinctive form of belief in disease deities, especially goddesses. Hindu society, rightly observed by David Arnold, did not regard all diseases alike. Some were seen as the consequence of personal sins, others the result of sorcery. But epidemic diseases, due to their scale and nature, and the general ineffectiveness of conventional medicine against them, were readily identified with the wrath of gods and cosmic disharmony.[15]

In fact, most of the disease goddesses were associated with a particular disease or ailment. In Bengal, as we will notice later, the principal deity was Sitala, the goddess of smallpox. The worship of Sitala was timed to coincide with the beginning of the smallpox season.[16] In deltaic Bengal, the specific cholera deity was called Ola Bibi by the Muslims and Olai Chandi by the Hindus.[17] Most interesting, however, is that the people did not interpret the malaria epidemic as a heavenly intervention. And consequently one finds no socially acclaimed folk deity for fever epidemic, except of course, the cult of *Jwarasura* prevalent among the lower classes of people. Though it has been argued that the fever-demon *Jwarasura* is present in the textualized 'classical' traditions,[18] it appears all the more certain from contemporary literary imagination that only the marginalized castes invoked *Jwarasura,* the fever demon, with the help of a Brahmin during malaria epidemics so that individuals and groups may recover.[19] *Jwarasura* was, in fact, never 'venerated', or 'worshipped' either like other disease deities for the simple reason that he was a 'demon' or *asura,* and not a 'deity' or *deva* as such. People worship devas or godhead out of reverence or fear; *asura* is invoked out of irreverence and disrespect. However, apart from the usual offerings of rice, sweets and fruits, goats were sacrificed in special instances for invoking *Jwarasura*. This being seems to have had no following among the upper-caste Hindus or well to do villagers. Nowhere in contemporary literary work nor in the official records could one find the reference to any widespread worship of this particular deity. In other words, this disease deity, unlike Sitala or Ola Bibi, never became a cult in the Bengali society. This provides an interesting exception to Arnold's general observation that almost all epidemics in India are associated with a particular disease deity.[20] And this calls for an explanation.

The explanation may be two-fold: the lack of indigenous immunity, and the exceptional geographical range of the disease. The causes of malarial fever, and quinine, its supposed herbal remedy, were not known to people till the advent of the twentieth century. Until then, people sought solace in the popular adage: that which cannot be ended should be mended, and which could not even be mended should be accommodated within the regular pattern of life.[21] Moreover, malarial fever might be remittent or intermittent; it abated at intervals. Its habit of periodic invasion provided the patient temporary respite from fever, and this helped reduce fear of it to some extent. Besides, this type of ailment was not unknown to the people. The affliction had accompanied them from cradle to coffin so that the disease had lost much of its shock element.

Secondly, epidemic malaria lost much of its edge due to its widespread spatial expansion. It traveled from Assam to Amritsar, from Bombay to Bengal. It was not an isolated ailment, afflicting a particular person or two. Particularly in Bengal, few districts could escape its ravages. The epidemic became almost a national phenomenon, visiting different regions simultaneously. Thus there prevailed an attitude of helpless resignation to conventional treatment and the epidemic was scarcely regarded as a sign of divine displeasure. Thus the worship or propitiation of a particular deity was ruled out.

On the contrary, people accused the colonial government of arbitrary interference with the river system in eastern India and consequently bringing about a virulent epidemic. Railway embankments, people argued, had converted large tracts of fertile land into perpetual swamps. Increasing numbers of embankments and railways reduced the chances of flood and inundations, and the reduced chances of river flood meant progressive loss of alluvial soil. The consequences, people argued, were generally two-fold. First, impeded drainage brought in its train large numbers of stagnant pools and marshes, breeding grounds for mosquitoes. And second, loss of fertility of the soil due to cessation of inundation had diminished the quantum of winter harvest, which meant starvation for a sizeable section of the rural masses.[22] The government, they complained, was in no mood to adopt effective measures to drain out rivers, particularly the Damodar.

Though the people lamented acts of omission, they protested against the rulers' acts of commission as well. Since many of the affected villages became particularly overgrown with tangled vegetation, government officials pleaded for the cutting of superfluous tree growth in the vicinity of the dwelling houses. So the government was well intentioned, but some of the over-zealous officials resorted to measures that virtually denuded the villages of their greenery. They chopped off branches, trimmed the shrubs short and worst of all, uprooted the larger trees, particularly the fruit bearing ones.

The high-handedness of the officials never went uncontested. People in the affected villages, particularly in Jessore district, were up in arms when government officials approached the villages with personnel to execute government orders. Entrusted with the unpleasant task of pulling down large trees, officials had to beat a hasty retreat.[23] They also encountered similar resistance from villagers when the government opened up *langarkhanas* or gruel kitchen in the immediate vicinity of the affected villages. When the government first introduced a relief-house at Burdwan, people opposed the system of giving cooked food on religious grounds, and sarcastically called it 'hotel'.[24] The Hindus certainly did not approve of the idea of cooked food being distributed; upper caste Hindus thought it preposterous to join the 'lowly' crowd of alms-receivers. The method of food distribution too added insult to injury. Each individual was furnished a printed ticket which entitled him or her to receive the stipulated amount of food relief. The recipients, it was admitted by the government officials, generally consisted of the lower caste people, mostly women and children of humble social status. Nevertheless, the Muslims seemed to have no inhibitions with cooked food, for at least at one depot at Nilpur they were very keen to receive the relief provided by the *langarkhana*.

Popular protest to 'public health' measures in terms of cutting of trees was thus noticeable. Opposition to the provision of food relief was also discernible. But opposition to government dispensaries was seldom noticed. The government opened up fever hospitals and fever dispensaries in the affected areas and provided quinine free of cost to the sick. People very often complained of the irregular supply of 'adulterated' quinine. Initially; the people seemed to be rather indifferent to such establishments, partly because the doctors in-charge were mostly unconcerned about treatment and irregular with their attendance.[25] Gradually, however, these institutions became popular, and an increasing number of patients, constituting a 'crowd', would collect at the outpatient departments. Statistical evidences from the hospitals in Calcutta as also in the amalgamated portion of its suburbs and in Howrah indicate as many as 2,08,073 visited hospitals in 1890.[26]

The intensity of malaria and the consequent hospitalization of the afflicted thus evoked little or no resistance from the masses. This stood in sharp contrast to the phenomenon of plague epidemic in Bombay. The physical examination of travellers and the residents of plague-struck towns and cities, hospitalization and segregation produced tremendous alarm, and people opposed the measures tooth and nail.[27] Because most of the inspecting doctors were male and European, their touch was considered either polluting or tantamount to sexual molestation, particularly when it involved the examination of women's necks, armpits, and thighs. David Arnold dwells on the repugnance of the oriental bodies to Western medicine for reasons religious and social, their loud protest against subjection to European doctors' examinations, and finally the mitigation of harsh government measures to placate the turbulent circumstances.[28] Nothing of this sort happened in Bengal when initially the Company and then the Crown resorted to a sort of social segregation by way of internment in hospitals and doling out Western medicine through public distribution systems such as dispensaries and post offices.

In the case of plague, people considered it as an expression of divine wrath. Many Indians saw the plague as a form of divine punishment, a visitation against which the use of Western medicine was bound to be either impious or ineffective.[29] But malaria, unlike the plague, was never interpreted as a divine intervention, far less a punishment by the heavenly hand. In collective imagination its virulence was associated with the baneful effects of British colonial policy. People argued that since the government had stirred up the disease, none but the government should take the responsibility of remedying the situation. Second, treatment of malarial fever with indigenous medicine proved ineffective. In most cases death was hastened by 'injudicious treatment from ignorant and unqualified native practitioners'.[30] Government officials believed that the dreadful mortality was due primarily to wrong treatment by the *kavirajas*, and people as such were increasingly losing faith in their treatment.[31] At the same time, the people admitted the efficacy of European medicine in sharp contrast to the tardy and uncertain cure under the native system. Therefore opposition to Western medicine was out of question.

Nevertheless, when the epidemic fever first struck Bengal countryside, people were filled with consternation. Gradually the fever travelled from one end of the district to the other, killing some and leaving others shattered in body and soul. People had never seen so many others die in their villages in so little time. Funeral processions with the ritual chanting of *Hari* stalked the countryside every day. The funeral pyres at the burning ghats were never put out; when the mortal remains of one person were reduced to ashes, those of another were ready for cremation. Elderly inhabitants confessed that they had never seen such a terrible visitation. From nearly every house the cry of lamentation could be heard.[32] Only when the government started sending out 'native doctors' trained in Western medicine and opened up charitable dispensaries, did public fear gradually subside.

Available evidence indicates that poor peasants refused to move out of their homesteads to healthier districts even in periods of high malaria mortality. Government officials could hardly induce the villagers to leave their family residence, the *bhadrasanbati*, in which were installed the family deities and guardians of the spirits of their ancestors.[33] Migration to healthier places, particularly to Calcutta, the capital city, was, however, common among the well-to-do families, especially among zamindars.

THE RURAL GENTRY

This brings us to the perception of the rural gentry. Unlike the plague, malaria was more severe in the countryside, where, not unlike Madras, little malarial operations had been taken up by the government.[34] So it is worthwhile to notice how the rural gentry reacted to the proliferation of the disease. The attitude of the better-off peasants, smaller zamindars, and other wealthy influentials to the fever epidemic was more or less practical, and the methods they evolved to circumvent it was pragmatic. They were not necessarily fatalist, and they did not worship the *jwarasura* either. They used to call on the village *vaidyas* or *kavirajas*, failing which they looked for better cures. They could visit the government dispensaries attended by local doctors practising Western medicine. At the last resort, they could leave the affected village for good and settle in the city. This was more or less common with the zamindars.[35]

When the government resorted to clearing forests in an attempt to destroy the habitat of mosquitoes, the better off and more influential 'natives' proved to be the most troublesome. They were opposed to such operations and tried by every means to incite the villagers to see to it that such measures were never carried out.[36] Conspicuous among them was the Manager of Prasanna Kumar Tagore's estate; government officials believed that it was solely due to his influence and advice that so much opposition was encountered. The government was compelled to abandon the work altogether.[37]

Other zamindars followed suit. In fact, the officials found it intriguing that when the sole object was 'how to do good to the sick and suffering', obstacles were 'raised in the most unexpected quarters'.[38] The administration expected the zamindars to play an active and more meaningful role. Even the people hoped that the zamindars would come forward with bountiful aid as relief for the *raiyats*. Should the peasants die, they argued, agriculture would suffer and revenue would fall.

There were of course some exceptions. The Maharaja of Burdwan and Jaykrishna Mukherjee of Uttarpara did come forward to help their subjects. To the Maharaja of Burdwan the government was indebted for almost entirely defraying the cost of fever dispensaries in Burdwan for several years. Besides, he made a large donation towards provisions of relief to the people who had suffered from the ravages of fever in his hometown, Burdwan.[39] Jaykrishna Mukherjee also persistently prodded the government to impress upon them the need for immediate action, and under his own supervision he had improved irrigation and the supply of drinking water in Dwarbasini.[40] His initiative and ingenuity in this respect attracted the interested attention of several other zamindars who had hitherto neglected the provision of clean drinking water to their raiyats.[41] He was greatly annoyed with the ineptitude of the government in handling the malaria question, and deplored the tendency of the government engineers to ignore the collective wisdom of local people based on observation and experience.

Nevertheless, the fact remains that most of the Bengal zamindars were apathetic to the general welfare of their raiyats and more so in matters of health and hygiene. Most of them had neither the inclination nor the means to undertake these works. This was partly because their income from land was increasingly dwindling under the revised government demands from time to time. In fact, most of the zamindars were contending with the grudge that in many cases they had not been allowed a margin of more than 5 per cent of the total revenue collection including establishment charges. Nor was there, they argued, any stipulation in the contract of the Permanent Settlement about irrigation work.[42]

If the response of the zamindars as a class was generally hostile and their reaction skin-deep, their scions seemed more concerned and some newly educated middle class gentry found the epidemic fever a cause to fight. Some major characters in the novels of Sarat Chandra Chattopadhyay, a litterateur well-known for his empathy for the unlettered and poor villagers, illustrate this point. Ramesh in *Pallisamaj*, Brindavan in *Panditmasay*, and Srikanta in *Srikanta* took up the cudgel to eradicate diseases like malaria, cholera and plague from the countryside. Some of the novels of Tarasankar Bandyopadhyay depict the concern of zamindars' sons about epidemics. In *Ganadevata*, for instance, Debu, a scion of a well to do family imbibed philanthropy and plunged headlong into movement for eradicating diseases from the ailing masses. Again, in *Dhatridevata*,

Shivanath is an archetypal member of the rural gentry, who proved to be a good samaritan, fighting epidemics and also social injustice.

The imaginative reconstruction of the novelists finds unwitting corroboration in local literary tracts of the period. Contemporary drama depicted the worry of the educated middle class about the sufferings caused by malaria. Realizing the attitude of the government, a number of public-spirited youth started anti-malaria societies to raise money to which a contribution by the government had been added. Zealous youth sought to organize Anti-Malaria Cooperative Societies in affected villages, which undertook various sanitary measures to improve public health. With the help of this fund, they organized branch societies in the malaria stricken villages of Bengal. Their modus operandi was similar to one adopted in the United States, as described in *American Journal of Medicine*.[43] The Central Cooperative Anti-Malarial Society, in particular had been completely successful in attaining its goals in two villages that they had targeted in 1921. They had been flooded with applications from hundreds of villages to do similar work in their respective villages, but were prevented from taking up the work due to financial constraints.

The Society approached the Rockefeller Foundation, New York with a request for assistance in its attempt to save forty-seven million people from the clutches of malaria. Dr. Gopal Chandra Chatterjee, the Secretary of the Society, requested that Dr. C.A. Bentley, Director of Public Health for Bengal, forward a letter of appeal to the Rockefeller Foundation.[44] In November 1922, Dr. Bentley forwarded the application request. The Society, he believed, was worthy of every possible assistance, for he had personal knowledge of a number of affiliated societies and was convinced that the movement was organized along the right lines. Bentley strongly supported the application for monetary help. Completely apart from anything else, the educational effort of the work was extremely valuable, he believed.[45] The International Health Board of the Rockefeller Foundation, however, empathized with the concern of the Society and appreciated the work it had undertaken, but it did not help with personnel or money. Its primary interest lay in the scientific research of malaria rather than the financial support for people in distress.

The work done by the anti-malarial societies can scarcely be questioned, but very often the volunteers of such societies used to come to blows with village bosses over the question of filling in vegetated ponds.[46] Undaunted, they continued their yeoman service and raised funds to reach out to a larger body of affected people with all possible help. The attitude of the educated middle class to the measures undertaken by the government had also generally been appreciative. They deplored the callous indifference of the people and exhorted them to stand by the government.

THE BENGALI INTELLIGENTSIA

Our analysis of the way the Bengali intelligentsia engaged with the problem depends heavily on literary evidence. Almost all important litterateurs of the period alluded to malaria. Sarat Chandra Chattopadhyay, for instance, found in malaria an appropriate cause for convincing his readers of the insanitary condition of the villages. A good number of characters in his novels suffered from malaria for weeks or even months together. To Sarat Chandra malaria was an inescapable destiny of rural life. He had seen regular festivities in the countryside seriously marred by malarial intervention. People suffered from attacks of malaria repeatedly without any respite in sight. It had been an inevitability ordained by *Jwarasura*.[47] At times, Sarat Chandra argues, people suffering prolonged spells of fever developed a stoic indifference to life. Shambhu Chatterjee in *Arakshaniya*, for instance, takes his dinner early in the evening and waits under his blanket for the onset of the fever. Sometimes, however, prolonged malaria used to provide a convenient tool for the disintegration of familial ties. Heroines of Sarat Chandra often found in the fever a convenient excuse for temporary separation from their estranged kith and kin.[48]

Sarat Chandra, it is true, depicted malaria as a deadly disease. But to him cholera was a greater killer, while plague was still deadlier.[49] Nevertheless, he was aware of the fatal impact and demoralizing influences of malaria. Village girls were often married off to malaria-stricken men eventually to end up as widows. These widows used to live out their lives in their paternal houses, abused evermore and absolutely uncared for.[50]

References to malaria are prolific in the novels of Bibhutibhusan Bandyopadhyay, who was aware that most of the villagers could not procure mosquito nets to protect themselves.[51] Once flourishing villages were depopulated. Critics argue that the supernatural setting of *Devjan*, one of the finest novels of Bibhutibhusan, was inspired by the malaria stricken countryside of Bengal.[52]

To Rabindranath Tagore, malaria was much more than a disease; it reflected in microcosm, he believed, many of our contemporary social evils. He was pained to see people fighting among themselves on petty matters instead of fighting out the epidemic. Immense good could have been done, Tagore argued, had the villagers joined anti-malaria societies in larger numbers.[53] He had nothing but praise for the organizers of those societies not only because they were fighting mosquitoes but particularly because they had overcome their idleness and started movements against social disunity.[54] The reason why the country had been repeatedly subjected to alien rule, he argued, can well be attributed not to our physical weakness but to our lack of organic social unity. National solidarity could well be promoted should people organize themselves to fight their own misery. Malaria had indeed offered them an opportunity towards this much-needed social unity and assertion of their national identity.[55]

Tagore was not unaware that malaria epidemics in Bengal were a function of colonial policy. He also shared the widespread popular belief that railway embankments, which obstructed the natural drainage system of the countryside, contributing to the proliferation of malaria, aggravated the epidemic. He traced the root of malaria to the rapacious greed of the colonial government. The country had been bled white by the British exploiters. The physical constitution of the people had also become weak as a result of recurrent flood, famine and fever. Nowhere in the world, lamented Tagore, could one find a people more indolent than the Bengali. Tagore was pained to see such a colossal loss of vitality, such listlessness, pessimism and despair writ large on the face of the people. Malaria, he argued, was largely responsible for this.[56]

THE CALCUTTA PRINT WORLD

The medical community in Calcutta was not particularly happy with the government's lack of initiative in incorporating new knowledge in its policy to control malaria. The *Indian Medical Journal* lamented in 1929 that the knowledge gained after the researches done by the School of Tropical Medicine, Calcutta had not been fully applied to fighting malaria.[57] An almost similar opinion was expressed in the *Indian Medical Gazette* in 1936 with special reference to Ross's discovery.[58] V.R. Muraleedharan narrates the experiences of the Madras Presidency in applying knowledge of the carrier of malaria, and concludes that what followed was in fact an increasing reliance on palliative measures of temporary value rather than a well-thought-out policy of complete eradication.[59]

Unfortunately Bengal was no exception. The response of its community was reflected in the print world. The reaction of the press was sharply critical. Almost all the journals published in Calcutta criticized the government for its half-hearted response to the epidemic, and in this regard perhaps the *Hindoo Patriot* stole the show. In 1872-3 it published a series of articles endorsing the views of Raja Digambar Mitra, a member of the Epidemic Commission, that the virulence of the epidemic was due to the obstruction of natural drainage. It sparked a lively debate with the government on this issue. The *Hindoo Patriot* cited copious examples to prove that the epidemic and its consequent mortality had been an outcome of British colonial policy.[60]

In the *Calcutta Review* too, fever figured quite frequently.[61] Major W.H. Gregg, Sanitary Commissioner for Bengal, wrote approvingly of the best arguments in Mitra's thesis and emphasized that the most important factor, which produced malarious conditions in the countryside, was impeded drainage. As an important government official Gregg was, however, a little cautious in his argument lest it implicate his masters. It was the inordinate humidity of the towns and villages and not the paddy fields and *jallas*, he argued, which contributed to the outbreak of severe malarial fever.[62] Gregg, however, did not expect the mofussil municipalities with their limited resources to accomplish as much in the direction of drainage as the Calcutta Corporation and other wealthy municipalities had done in a short period of time.

One important issue in all such publications was that the authors were frequently the beneficiaries of British rule. They were either the employees of the Public Health Department or represented the landed interest. As loyalists, they were somewhat cautious in their criticism of the government. Girindra Krishna Mitra is a case in point. An employee of the Public Health Department, Mitra published a treatise on malaria under government patronage.[63] He rightly observed that the question should not be considered in isolation from the socio-economic milieu. But he erred in judgement by laying the blame entirely on his own people. Mitra argued that the villagers were increasingly becoming careless in their sanitary habits. He regretted the migration of the rich to healthier towns, leaving their native villages in shambles. He lamented their lack of initiative in rural development. But he consciously underplayed the responsibility of the Raj.[64]

Intellectuals with no axe to grind were more forthright in their criticism. For instance, in a monograph on malaria Raj Krishna Mandal, showed how the progressive state of pauperization of the rural masses drove them to epidemic diseases.[65] He emphasized the phenomenal changes in crop pattern leading to the cultivation of cash crops, such as sugar and cotton. From the cultivation of sugar cane and cotton the peasants now shifted to that of jute, a product that exhibited enormous export potential. Large increases in the cultivation of jute tended to bring about a corresponding decrease in the cultivation of paddy, which had hitherto met the basic subsistence needs of a Bengali household. Thus followed a protracted period of undernourishment, which Mandal argued, was one of the causes of epidemic.[66]

Likewise, most of the vernacular journals published in Calcutta categorically condemned British policy. *Swasthya*, a monthly magazine on health and hygiene, argued that stagnant pools were the principal source of mosquito breeding. It looked on to the rajas, maharajas, zamindars, mahajans and particularly the government, and insisted that they come forward to help the afflicted with monetary support and medical relief.[67] It condemned what it called the 'double standard' followed by the government. When the mortality from malaria reached its peak, the government deputed *chowkidars* in the villages and police in the towns to collect information on the exact number of deaths. The government, it argued, thus spent a fortune on preparing the death rolls but only a pittance on prevention. In England, public health policies had dramatic effects in the control and elimination of epidemic diseases, but in India, it complained, the government had shown the least concern.[68] The *Swasthya* also bemoaned the migration of wealthy people to the towns. But it did not always hold the migrant errant. It argued that so long as the villages suffered a scarcity of pure drinking water, people fell victim to the vicious cycle of malaria and malnutrition, and nobody then could desist emigrating to the towns.

The *Swasthya* suggested certain remedial measures to ameliorate the situation.[69] (i) The municipalities and cantonments should prepare a list of overgrown village ponds in their respective localities through their own doctors, and arrange to fill them up, (ii) legal provisions should be made so that such breeding grounds of mosquitoes were not dug out in future, (iii) brickfields should be shifted outside the municipal areas, (iv) the digging out of earth from under ditch embankments for repairing village roads should be strictly prohibited, (v) the practice of digging out ditches for laying railroads should be discarded if they ran through villages. In fact *Swasthya*'s attack was two-pronged: (a) it tried hard to impress upon the ignorant masses the essentials of good hygiene and sanitation and (b) it criticized the government in no uncertain terms for its apathy and neglect of village welfare.

The *Krishak* was still more vociferous in its indictment of the educated urban middle class for its apathy and neglect towards disease and devastation. The urban intelligentsia, it argued, had raised voices against British rule and clamoured for greater self-government, *swadeshi*, labour union, tenancy legislation and sundry other matters. But it never bothered about the health of peasants who constituted the bedrock of zamindar prosperity. The peasants had no organization of their own, and they had looked up to the educated middle class for guidance. But if the middle class betrayed their cause, *Krishak* wondered, who would the peasants look up to?[70]

In a strongly worded article, *Krishak* criticized the government for having promised Rs 4,48,000 for the public works department, and only Rs 78,000 for agriculture. Malaria, which proved to be the chief constraint on public health, received no budgetary support at all. Lord Ronaldshay was reportedly dismayed at the amazing number of deaths from malaria, but the *Krishak* wondered about how much the government decided to spend for the eradication of such an evil. The *Krishak* asked sarcastically: 'Could not the government spend for the eradication of malaria the amount of Rs 2,46,000 which it spent for the annual upkeep for three ministers and one additional member on the Executive Council?'[71]

The *Krishak* contended that malaria in its epidemic form was not peculiar to Bengal, nor was it an unknown ailment in the world. Countries elsewhere witnessed appalling instances of malaria ravages. Britain too had experienced extensive marsh fever which took an alarming human toll. But all those countries had successfully rooted it out through a persistent policy of health care. In Bengal, it persisted owing to the abject poverty of its people and blatant neglect of rural sanitation. And the government itself was responsible for the situation.[72] By way of explanation, the *Krishak* argued that the villages in Bengal no longer drew the middle class. Gone were the days, *Krishak* regretted, when those on service in the towns used to visit their village homes during the festival days and spend lavishly on social festivities. They used to feed the hungry, repair roads, dig or clean ponds, and arbitrate village disputes. Festive days over, they would return to their place of work only to revisit it in the next year. But unfortunately, this age-old practice was no longer in vogue. Villages were now reduced to a state of abysmal poverty and squalor.[73]

Gradually, however, the vernacular press started driving home the Biblical adage: 'God helps those who help themselves.' By 1922 the *Swasthya* began to argue that it was an act of utter foolishness to depend so much on the government; people themselves should find their way out. The government had spent lakhs of rupees on the 'foreign doctors' for their 'fruitless exercise' in research and experiments on malaria. People had had enough of such experiments; what they really needed was some tangible work. Reports piled upon reports while the Bengalis as a race were dying out in alarming numbers. The reports should stay put in shelves and preserved properly, for, the *Swasthya* argued, they would ignite the funeral pyres of the victims![74]

The malarial fever, its intensity and high rate of mortality, evoked widespread interest and concern in the reputed medical journals too. Doctors, Indian as well as European, joined the debates on the issue of its causation, dissemination and killing properties. Some noted physicians of Bengal who practised medicine and taught in the medical colleges contributed scholarly articles to all important medical journals of the time. Dr. U.N. Brahmachari, along with Dr. C.A. Gill, Dr. S.R. Christopher and others addressed the issue of causal relations between density of population and malaria in the *Indian Journal of Medical Research*.[75] Dr. Brahmachari wrote another article in the *Indian Medical Gazette* on the nature of the epidemic fever in Lower Bengal.[76] Dr. Sarasi Lal Sarkar investigated the incidence of malaria in the town of Arambag, Hooghly district, and brought to light some startling revelations.[77] Dr. B.B. Brahmachari explored the economic problems involved in the 'fever of Bengal', and showed that the disease, while it spread from *thana* to *thana* and from district to district, found the people labouring under a very serious strain caused by (a) the density of population, (b) shifting of traffic from the existing old centres, (c) decay of industries, and (d) deterioration of agriculture.[78]

In fact, the impact of malarial fever on industry caught the interested attention of many an Indian doctor. Dr. R.N. Sen, for instance, narrated how malaria caused havoc in different coalfields of India.[79] Dr. K.L. Chowdhury looked into the impact of malaria on the supply of labour to the tea estates of Bengal Dooars. Dr. Chowdhury and Dr. MacDonald attempted an estimate of financial loss to tea gardens caused by the incidence of recurrent malarial fever.[80] Dr. Gupta, Dr. Das and Dr. Majumdar made a joint survey on malaria ravages in the tea estates of Goalpara district, Assam, and suggested some significant remedial measures.[81]

However, the issue, which aroused widest interest in all medical journals, was the relation between malaria and pregnancy. Dr. S.M. Das analysed the causes of abortion, and concluded that malaria exerted a very unfavorable influence on pregnancy.[82] He deplored the indifference of husbands to the plight of their wives on account of too frequent conception and exposure to malarial infection. Similarly the question of administration of quinine during pregnancy was a much talked about concern. Some doctors apprehended that quinine induced abortion. Thereupon, practising physicians all over the province started sending in letters and articles narrating their own practicing experiences. One Dr. Chandra Kumar Datta from Noagaon wrote a letter to the editor of the *Indian Medical Gazette* stating that his experience had proved to the contrary.[83] Dr. Upendranath Chakraborty from Tangail supported Dr. Datta, and contended that 'quinine is almost the only drug on which we have to fall back to stop malarial fever after all other drugs have been tried, and no harm is done even if the women be pregnant'.[84] European doctors too joined the controversy, and after much debated deliberations they concluded in favour of quinine administration.[85]

Finally, some doctors evinced an active interest in the bacteriology of the disease. They were concerned with the spread of malaria and kalazar, and engaged in this research. They categorically denounced the role of government and condemned its lamentable lack of concern in public health and sanitation. Dr. Ganapati Panja, an Assistant Professor of Bacteriology, School of Tropical Medicine, Calcutta argued in 1923 that the preventive measures for malaria could be theoretically well planned as the etiology of the disease was well known. But considering the then economic condition of the people and the attitude of the government, Dr. Panja pleaded, it would be useless to suggest expensive measures.[86]

Nevertheless, Dr. Panja considered education was the only panacea of all evils. First of all, he argued, people should be educated. 'We cannot help them if they do not help themselves', pleaded Dr. Panja. They should identify the causes of the spread of malaria and try to remove them. 'Students of village schools,' he argued, 'should be particularly educated.' Education on public health should be imparted through lantern lectures, and suitable models of malaria-stricken villages and mosquito-breeding places should be exhibited to the villagers. Students should be advised to form volunteer batches whose duty would be to look after the sanitation of their villages in spare time.[87]

In fact, following this line of arguments, Dr. Bentley, perhaps the lone supporter of malaria eradication in Bengal, contrary to the general lackadaisical attitude of the government, requested that the Rockefeller Foundation present the Bengal Public Health Department a gift of the film it had prepared for illustrating the natural history of malaria and the preventive methods against that disease.[88] Dr. Bentley hoped that it would be shown all over the province by the departmental publicity bureau.

Dr. Panja, however, was concerned with how to keep mosquitoes away. The plan of destroying mosquito larvae he considered useless, for he argued, experience had shown that nobody would take the trouble of carrying it out. But if some appropriate and cheap pyrethroid insecticides or some other was found out and placed at the disposal of local grocers, people could easily buy it and fumigate their house with it every evening. Dr Panja considered the use of mosquito nets impracticable, as it was not possible for all the villagers to procure them.[89]

In an article published in the *Indian Medical Records*, Dr. Amulya Chandra Mitra expressed concern over the rate of deaths from malaria. He was pained to notice that epidemic malaria had virtually devastated all the districts of Lower Bengal.[90] The census reports of various districts revealed the awful state of health, and if such state of affairs was allowed to drift further, the condition of the people, he apprehended, would be alarming. The industries, he observed, were dying out for want of workmen, and even cultivation would have to be abandoned if labourers were not recruited from other provinces. The overall condition of the country was going from bad to worse, and it required 'urgent measures to improve the pitiable sanitary condition of the province and save the millions of people from the scourge of malaria'.[91]

Dr. Mitra suggested some practical measures, which he hoped, would bring in the desired results. Attempts should be made to organize sanitary boards in villages with the active help of young people trained and assisted by local medical men and guided by health and sanitation officers of the province. The officers would educate the people through demonstrations, pamphlets and lantern lectures, and advise them about precautionary measures against the outbreak of epidemic disease. Dr. Mitra applauded the works of the Cooperative Anti-Malarial Dispensary, an organization that owed its origin to the indefatigable energy of Dr. Gopal Chandra Chatterjee. It had done immense good to the people.

Dr. Mitra did not repose much faith on the government initiative. He pinned his hope on the District Boards. The District Boards, he pleaded, should come forward with adequate funds to sustain the activities of the Cooperative Anti-Malarial Dispensary. The Co-operatives not only helped members but also others to be treated by such organizations. While it was not possible for the Boards to open large dispensaries adequately equipped to meet the demands of the people, it should render all possible help to the Cooperative Dispensaries, especially with money.[92]

VILLAGE MEDICAL PRACTITIONERS

The response and reaction of the indigenous medical practitioners based at villages, was just the opposite. The really competent *kavirajas* being always very few, the quacks had had their heyday in the epidemic years. Most of the indigenous doctors, notably some of the village *kavirajas*, found in malaria a scope for increasing their income. Worse still, the War witnessed a proliferation of adulterated quinine, the only drug which seemed to have some remedial properties. There were some conscientious practitioners who seemed to have been aware of their medical ethics and social responsibility. They sincerely attempted to cure the patients. There were still some more who wrote medical treatises on malaria and other diseases, and called their countrymen to follow the general rules of personal hygiene and community health.

Generally, the village *kavirajas* used to keep indigenously prepared 'fever mixture', and administered it on payment of a Re. 1 or 2 when patients approached them.[93] By 1908, the cost of quinine rose to Rs. 25 per phial, and no quack doctors ever administered the required dose to a patient. For instance, they used to charge money for 50 grains, but actually offered only 10 grains.[94] As a result, fever relapsed immediately after two or three days, and the patients were forced to visit the doctor again, and the same under dose of quinine was repeated in a long drawn process until the patients breathed their last. On an average a person could catch fever five to eight times in a lifetime, and on severe circumstances, he had to spend at least Rs. 10-20 for treatment.[95] The doctor's visits on every alternative day cost a family no less than Rs. 4. Doctors of the Anti-Malaria Society, however, charged much less than that. Available information indicates that it was not more than Re. 1 per visit.[96] Visit to a patient living in a distant place by a qualified doctor, cost an enormous amount, sometimes around Rs. 150.[97] The amount included the cost of medicine, the doctor's fee, and the charge of palanquin bearers, who brought the doctor.

Nevertheless, adulteration of quinine both by doctors and producers became rampant. It was in fact widely practised even before the First World War. During the War its prices increased so much that indiscriminate adulteration continued on a large scale. In the mofussil area one ounce of quinine (Quinine, Herrings and Co.) sold for Rs. 2-8 in 1916, while the same quantity of adulterated quinine could be had for just Rs. 1-2. The label of the latter was identical with that of the former, only the block being a bit fainter. This quinine effervesced on the addition of any acid. Possibly it contained in it magnesium carbonate and sodium bicarbonate.[98]

Quacks could not check the temptation to dispense this cheap quinine to their patients. The result was obvious. When they failed to check the shooting fever, they blamed the allopath or the quinine. The unscrupulous shopkeepers also tried to pass the adulterated quinine for the genuine one and misled the gullible patients.[99] Quinine thus became unpopular in Bengal so much so that at times the physicians had to use their own symbols and signs for it in their prescription and not to mention it by name to their patients. Conscientious doctors therefore pleaded that that adulteration of such a drug, indispensably necessary for the malaria-stricken population, should not be allowed to continue with impunity.[100]

Instances of treatment of malaria by quacks with adulterated quinine are prolific in contemporary Bengali novels and short stories. In *Ramer Sumati*, Sarat Chandra narrates a sordid tale of how a country quack sought to prolong the treatment of malaria in the case of a young housewife. He presumably administered adulterated quinine, and the patient continued to suffer. Increasing deterioration of health worried her kith and kin, and the brother-in-law, a short-tempered boy, stormed into the quack's chamber. He abused the quack in the presence of other malaria patients there; the quack sought to mobilize the latter to teach the boy a lesson, but they backed out because they too had passed through the same trauma of adulterated quinine.[101]

In fact, the quacks took greedy advantage of the malaria season. Right from daybreak, patients from far-off places used to gather at the doctor's chamber. Crowds of patients continued to collect till dusk, and the doctors complained that they missed their meals. The doctors used to compete with one another for the size of the crowd they could pull. In the patients' perception the caste to which the doctor belonged was also important. Generally speaking, the patients preferred a Brahmin. Even if a quack they preferred to 'die at his hand'. This attitude might have been governed by social circumstances; the doctors were aware of this phenomenon, and they even boasted of it.[102] Quacks of non-malarious zones often migrated to epidemic zones. These migrations were prompted by anticipation of finding larger numbers of patients and still larger amount of money. The migratory period sometimes lasted for five months; the doctors used to earn a fabulous sum during their sojourn.[103]

Malaria boosted up the business of others too. Recurrent malaria deaths in quick succession in a family were sometimes explained by ignorant lower-class people as the expressive wrath of the *Jwarasura*, the fever demon. Unable to cure the fever with adulterated or under-dose quinine, they resorted to providential remedy. Those who administered divine therapy sold talismans, amulets, magic rings, and promised the afflicted an early cure.[104]

Indigenous country physicians were not always governed by financial considerations. One can show that there were instances of a few doctors who took care to cure the disease in right earnest. Jagabandhu of Surma Valley is a case in point. He had been educated at the Medical College, but he failed to pass the examinations. Thereupon, he started practising on his own. His little 'dispensary' was clean, tidy and well-equipped; he seemed to take a genuine interest in his profession. He inveighed against the 'native doctors', the quacks, who used to starve their patients, adulterated quinine with soda, and gave opium. He usually gave his patients pigeon soup, sago, stimulants and quinine. When Dr. Smith, the Sanitary Commissioner of Bengal, visited his dispensary in January 1870 he told him: 'I am a plucked man but try to do my best'. Dr. Smith was impressed by his sincerity and remarked that 'really he was doing very useful service to the people'.[105]

A few tentative conclusions can be drawn from all of these. Response to the malaria epidemic from different sections of society followed neither a homogeneous nor a unilinear trajectory. Differences stemmed from their widely differential perceptions of the epidemic. Generally speaking, the response of the masses was one of stoic acceptance, and not a dreaded repulsion. There is, however, nothing new in this kind of attitude, for an epidemic such as this had never been looked upon as a frightening disease in Bengal. It has all along been depicted in Bengali folklore as a living entity, often described as a cute belle who delights in frequenting village huts, and although often despised by the wealthy landlord, is invited in by the impoverished peasant families.

On a general note, reaction to the prophylactic intervention and public health measures became rigid over a period of time. It followed more or less a uniform pattern everywhere with marginal variations. Almost all sections of society looked at the colonial government and the policies with suspicion and disbelief. The accusation was sometimes violent, sometimes muted. Reputed doctors held the rulers responsible for the epidemic, while blaming the people for their apathy to personal health and hygiene. They constructed the epidemic from a purely medical perception, which was an error in judgement. They could have realized that the struggle against disease must begin with a war against bad government.

NOTES

1. *Atharva Veda* (tr. R.T.H. Griffith, *The Hymns of the Atharva Veda*, Varanasi: Chowkhamba Sanskrit Series Office, 1968, vol. 1, Book 1, Hymn 25, no. 4, p. 31.

2. Ibid. 1. 25. 1, pp. 29-30, IX. 8. 6, p. 456.

3. Ibid. V. 22. 5, p. 224.

4. Ibid. For 'fever' the word 'takman' is used. Takman, derived from 'tak', to fly at, to pounce upon, meaning that which suddenly seizes, is explained in the *St. Petersburg Dictionary* as a kind of disease, accompanied by eruption on the skin. But a comparison of the passage of the *Atharva Veda* in which the word occurs is sufficient to prove that fever in its many varieties, especially malarial fever, is the disease that is intended. The word does not occur in the *Rig Veda*, nor is it found in works later than the *Atharva Veda*. The word 'mashaka', i.e. mosquito is found in the latter, VII. 56.2, and its bites, VII. 56.3, pp. 353-4. The word 'kusala' occurs in VII. 6.10, p. 405, and bloody mouth in VII. 6.12.

5. Ibid., VIII. 6.11.

6. *Charaka*, C. I. 3. 34, 3.67.

7. *Susruta* V. 39.69, 70.71.

8. The medieval town of Gour was devastated in the sixteenth century, around 1575, probably by malaria. Aniruddha Ray, *Madhyajuge Bharatiya Shahar*, Kolkata: Ananda Publishers, 1999, p. 144.

9. David Arnold, 'Disease, Resistance and India's Ecological frontier, 1770-1947', in *Issues in Modern Indian History*, ed. Biswamoy Pati, Mumbai: Popular Prakashan, 2000, pp. 1-22.

10. Ira Klein, 'Malaria and Mortality in Bengal, 1840-1921', *Indian Economic and Social History Review*, 9, no. 2, 1972, pp. 132-60.

11. V.R. Muraleedharan, 'Malady in Madras: The Colonial Government's Response to Malaria in the Early Twentieth Century', in Deepak Kumar (ed.), *Science and Empire: Essays in Indian Context*, New Delhi: Anamika, 2001.

12. Rohan Deb Roy, '"An Unseen Awful Visitant": The Making of Burdwan Fever', *Economic and Political Weekly*, 43, nos. 12-13, 2008, pp. 62-70.

13. Ihtesham Kazi, 'Environmental Factors Contributing to Malaria in Colonial Bengal', in *Disease and Medicine in India: A Historical Overview*, ed. Deepak Kumar, New Delhi: Tulika, 2001; Sujata Mukherjee, 'Environmental Thoughts and Malaria in Colonial Bengal: A Study in Social Response'. *Economic and Political Weekly*, 43, nos. 13-14, 2008, pp. 54-61.

14. Rohan Deb Roy, 'Mal-areas of Health: Dispersed Histories of a Diagnostic Category', *Economic and Political Weekly*, 42, no. 2, 13 January 2007.

15. David Arnold, 'Cholera and Colonialism in British India', *Past and Present*, no. 113, November 1986, pp. 129-31.
16. R.W. Nicholas, 'The Goddess Sitala and Epidemic Smallpox in Bengal', *Journal of Asian Studies*, vol. XLI, no.1, November 1981, pp. 21-44.
17. Gopendra Krishna Basu, *Banglar Loukik Devata,* Calcutta: Ananda Publishers, 1964, pp. 25-8, 195-8; Samaren Roy, *The Bengalees: Glimpses of History and Culture*, New Delhi: Allied Publishers, 1996, p. 22.
18. Projit Bihari Mukharji, 'In-Disciplining Jwarasur: The Folk/Classical and Transmateriality of Fevers in Colonial Bengal', *Indian Economic and Social History Review*, 50: 3, 2013, 261-88. David Hardiman and Projit Bihari Mukharji, 'Subaltern Therapeutics in Indian Medical History', *Wellcome History*, 9 July 2014.
19. L.S.S. O'Malley, and Manmohan Chakraborty, *Bengal District Gazetteers, Howrah*, Calcutta: Bengal Secretariat Book Depot, 1909, p. 46.
20. Arnold, 'Cholera and Colonialism in British India', op. cit., p. 129.
21. Instances of such attitude of helpless resignation can be found in abundance in the novels of Sarat Chandra Chattopadhyay, the most popular being *Pallisamaj*, p. 140; *Panditmasay*, p. 94; *Datta*, p. 794. All these references are from *Sulabh Sarat Samagra*, Calcutta: Ananda Publishers, 1989.
22. Editorial comments of *Gramer Dak*, a bimonthly journal of Howrah Zilla Krishak O Hitakari Samati, Ashar-Sravan, 1334, pp. 20-1.
23. G.S. Hills, on Special Duty, to the Hon'ble A. Eden, Secy to the Government of Bengal (hereafter GOB). Proceedings (hereafter Progs) of the GOB, General Department, August 1863, no. 20, p. 41. Microfilm Roll No. P 170, vol. 8. West Bengal State Archives (hereafter WBSA).
24. C.T. Buckland, Officiating Commissioner of the Burdwan Division, to the Secy to the GOB. (No. 80T dated Burdwan, 7 February 1870). Progs of the GOB, Medical Department, Sanitation Branch, February 1870, no. 30, pp. 19-20. WBSA.
25. Vide Dr. Thompson's report on the general result of the working of each dispensary in Hooghly. Progs of the GOB, Political Department, Sanitation Branch, May 1870, no. 13, pp. 9-15. WBSA.
26. A. Hilson, Report on the Calcutta Medical Institutions for the year 1890. Progs of the GOB, Municipal Department, Medical Branch, June 1891, para 4, p. 155. WBSA.
27. Rajnarayan Chandavarkar, 'Plague Panic and Epidemic Politics in India, 1896-1914', in Terence Ranger and Paul Slack (eds.), *Epidemics and Ideas*, Cambridge: Cambridge University Press, 1992, pp. 203-40.
28. David Arnold, 'Touching the Body: Perspectives on the Indian Plague, 1896-1900', in Ranajit Guha and Gayatri Chakraborty Spivak (eds.), *Selected Subaltern Studies*, New York and Oxford: Oxford University Press, 1988, pp. 397-401.

29. Ibid., 'Plague: Assault on the Body', *Colonizing the Body: State Medicine and Epidemic Disease in Nineteenth-Century India*, Berkeley, Los Angeles, London: University of California Press, 1993, pp. 236-9.
30. Report of Dr. J. Elliot. Progs of the GOB, General Department, Sanitation Branch, March 1863, no. 108, p. 69, WBSA.
31. Report of Dr. R.F. Thompson. Progs of the GOB, General Department, Sanitation Branch, June 1868, no. 14, p. 19. WBSA.
32. Rev. Lal Behari Dey, *Bengal Peasant Life*, Calcutta: Firma K.L. Mukhopadhyay, 1969 (rpt.), p. 250.
33. Report of R.C. Mukherjee, condensed in Pillow's Report dated 24 March 1878. Vide Crawford, D.G., *Hughli Medical Gazetteer*, Calcutta: Bengal Secretariat Press, 1903, p. 138.
34. V.R. Muraleedharan, 'Malady in Madras: The Colonial Government's Response to Malaria in the Early Twentieth Century', in Deepak Kumar ed., *Science and Empire*, New Delhi: Anamika, 1991, p. 103.
35. Gurucharan Rakshit, 'Pallir Durdasa', in *Krishak,* vol. XXII, 1328 (Bengali Year), no. 5, p. 121.
36. G.S. Hill to A. Eden. Progs of the GOB, General Department, August 1863. Microfilm Roll No. P. 170, vol. 8, no. 41, p. 20. WBSA.
37. Ibid., p. 21
38. C.T. Buckland to Secy to the GOB to the GOB, Judicial Department. Progs of the GOB, May 1870, Political Department. Sanitation Branch, para 24, p. 24. WBSA.
39. Ibid., p. 25.
40. *Indian Daily News*, 9 March 1868.
41. Nilmoni Mukherjee, *A Bengal Zamindar: Jaykrishna Mukherjee of Uttarpara and His Times, 1809-1888*, Calcutta: K.P. Bagchi & Co., 1975, p. 282.
42. *Report of the 16th Annual General Meeting of British Indian Association, February 27, 1868, Indian Daily News,* 9 March 1868.
43. *American Journal of Medicine*, 8 November 1919.
44. From G.C. Chatterjee, The Central Cooperative Anti-Malarial Society Ltd, through the Director of Public Health, GOB, to the General Director International Health Board (IHB), Rockefeller Institute, New York, Calcutta, 28 August 1922, Rockefeller Foundation (RF), Box 14, Folder 2565, Record Group (RG) 5, Series 1.2 1922, Rockefeller Archive Center (RAC).
45. From C.A. Bentley, Director of Public Health, to the General Director, IHB, Rockefeller Institution, New York, 7 November 1922, RF, IHB of the RF, Box 145, Folder, 2565, RG 5, Series 1.2, 1922, RAC.
46. Haridhan Bandyopadhyay, *Banglar Shatru*, Sodpur: no pub., 1924, p. 16.
47. 'Bipradas', *Sulabh Sarat Samagra*, Calcutta: Ananda Publishers, 1989, p. 1428.

48. 'Mejdidi', ibid., p. 1616.

49. For an informed depiction of cholera devastation, one can look up 'Panditmasay', and for plague 'Srikanta, Part II'. *Sulabh Sarat Samagra.*

50. 'Panditmasay'. *Sulabh Sarat Samagra.*

51. 'Bipiner Sangsar', in *Bibhuti Rachanabali*, Calcutta, 1398 (Bengali Year).

52. 'Introduction' to *Bibhuti Rachanabali*, vol. VIII, p. 9.

53. Rabindranath Tagore, 'Malaria', a lecture delivered in the Anti-Malaria Society. First published in *Bangabani,* Jyoistha, 1331 BS. Incorporated in the *Rabindra Rachanabali*, vol. 14, Calcutta: Viswa Bharati, 1398, pp. 390-1.

54. Ibid., 'Samabaye Malaria Nibaran'. Address delivered in the Anti-Malaria Society, organized by Dr. Gopal Chandra Chattopadhyay. First published in the *Sanghati*, Bhadra, 1330 (29 August 1923). Incorporated in the *Rabindra Rachanabali*, vol. XIV, Calcutta: Viswa Bharati, 1398 BS, p. 388.

55. Ibid., p. 389.

56. 'Malaria', op. cit., p. 390.

57. 'Progress of Research in India', editorial, *Indian Medical Journal*, vol. 24, January 1928, 1929.

58. G.G. Molly, 'Medical Research in India', *Indian Medical Gazette*, vol. 71, January 1936, p. 42. Also vide Muraleedharan, op. cit., p. 111.

59. Muraleedharan, op. cit., p. 111.

60. 'The Epidemic Fever in Bengal'. Reprint from the *Hindoo Patriot*, 1872-3, pp. 28-34.

61. A few examples: *Mofussil Records of Bengal*, vol. 54, no. 108, 1872, pp. 204-22; *Decay of Villages in Bengal*, vol. 124, no. 249, 1907, pp. 394-408; *Raiyat in Bengal*, vol. 34, no. 68, 1860, pp. 240-50; *Embankment of Rivers in Bengal*, vol. 8, no. 16, 1877, pp. 329-43; *Malarial Fever in Bengal*, vol. 88, no. 176, 1889, pp. 378-84.

62. W.H. Gregg, 'Malarial Fever in Bengal', *The Calcutta Review*, no. 176, April 1889, p. 381.

63. Girindra Krishna Mitra, *Malaria o Kalajwarer Pratikar Samasyer Parikalpana*, Calcutta: no pub., 1924.

64. Ibid., pp. 1-7.

65. Raj Krishna Mandal, *Bange Malaria*, Calcutta: no pub., 1908, pp. 22-7.

66. Ibid.

67. *Swasthya*, ed. Durgadas Gupta, vol. 2, no. 9, *Paus* 1305 BS, p. 211.

68. *Swasthya*, vol. 2, no. 10, Magh 1305 BS, pp. 218-19.

69. Ibid., vol. 5, nos. 8 & 9, 1308 BS, pp. 275-6.

70. *Krishak*, Aswin-Kartick 1327 BS, vol. 21, nos. 6 & 7, pp. 191-2.

71. Ibid., Falgun-Chaitra 1327 BS, vol. 21, nos. 11 & 12, p. 338.

72. Ibid., vol. 22, Bhadra, no. 5, 1328 BS, p. 121.

73. Ibid., p. 122.

74. Gopal Chandra Chattopadhyay, 'Banglar Sochaniya Abastha o Tar Caran', *Swasthya*, ed. Brajendranath Ganguly, vol. I, no. I, 1328 BS, p. 4.

75. *Indian Journal of Medical Research, Supplement*, vol. 1, 1914, pp. 178-80.

76. *Indian Medical Gazette*, vol. XLVI, no. 9, 1911, pp. 340-3.

77. Ibid., vol. XLVIII, no. 9, 1913, pp. 343-6.

78. *Indian Medical Records*, vol. XLVIII, no. 8, August 1923, pp. 216-18.

79. *Indian Journal of Malariology*, vol. XVII, nos. 2&3, June-Sept. 1963, pp. 157-74.

80. *Records of the Malaria Survey of India*, vol. 2, no. I, March 1931, pp. 112-18.

81. Ibid., vol. III, no. 2, December 1932, pp. 253-60.

82. Sundari Mohan Das, 'Malaria and Abortion', in *Calcutta Medical Journal*, vol. XVIII, 1923, pp. 342-52.

83. *Indian Medical Gazette*, vol. 44, Feb. 1909, p. 78.

84. Ibid.

85. For more details on the controversy one may look up: A.C. Basu, 'Quinine in Pregnancy', *Indian Medical Gazette*, vol. XLVIII, no. 3, p. 113; A.G. Newel, 'Quinine and Pregnancy', *Indian Medical Gazette*, vol. XLIII, no. 7, p. 274; Bhupal Singh, 'Quinine and Pregnancy', *Indian Medical Gazette*, vol. XLIV, no. 2, p. 78; Editorial Comment, 'A Note on the Administration of Quinine in Case of Fever during Pregnancy', *Indian Medical Gazette*, vol. VLV, no. 4, pp. 135-7; Editorial Note, 'Treatment of Malaria in Pregnancy', *Indian Medical Gazette*, vol. IV, no. 11, p. 438.

86. Ganapati Panja, 'Prevention of Malaria and Kalazar in Bengal', *Indian Medical Records*, Sept. 1923, p. 244.

87. Ibid.

88. From C.A. Bentley, Director of Public Health, to the General Director, International Health Board, Rockefeller Foundation, New York, Calcutta, 19 June 1924, Box 201, RG 5, Series 1.2, Bengal 466, Folder 2560, RF, RAC.

89. Ibid.

90. Amulya Mitra, 'Malaria and Kalazar, and Measures to be Adopted by the District Boards towards their Mitigation', *Indian Medical Records*, August 1923, pp. 234-5.

91. Ibid., p. 234.

92. Ibid., p. 235.

93. Rajkrishna Mandal, *Malariar Karan o Pratikar*, Calcutta: no pub., 1908, p. 56.

94. Ibid., p. 56.

95. Ibid.

96. Bandyopadhyay, op. cit., p. 25.

97. Mandal, op. cit., p. 55.
98. *Indian Medical Gazette*, vol. LI, no. 12, 1916, p. 474.
99. Ibid.
100. Jnanendranath Datta's letter to the Editor, *Indian Medical Gazette*, vol. LI, no. 12, 1916, p. 474.
101. Sarat Chandra Chattopadhyay, 'Ramer Sumati', in *Sulabh Sarat Samagra*, pp. 1572 f.
102. Bibhuti Bhusan Bandyopadhyay, 'Bipiner Sansar', in *Manmayuri*, Calcutta: Sahityam, p. 132.
103. Ibid., p. 133.
104. Ray Karali Charan, *Bange Malaria*, Vasantapur: no pub., 1917, p. 42.
105. D.B. Smith to W.H. Ryland, Officiating Asst. Secy. to GOB (no. 147 dated Calcutta 29 January 1870). Progs of the GOB. Pol. Dept. San Br. Feb. 1870, no. 16, p. 14, WBSA.

CHAPTER 2

Cholera

Ever since Asa Briggs had issued a call, some more than fifty years ago, for further research into the social history of epidemics, historical monographs on cholera started proliferating all over the world.[1] Rosenberg produced his magnificent work on cholera epidemic in the United States, McGrew on Russia, Durey on Britain, Delaporte on France, and many others on many more countries undertook similar studies. They enquired how societies coped with, reacted to, and interpreted short-term epidemic crises.[2] So far as India is concerned, our knowledge on the relations between cholera and colonialism has been enriched by David Arnold.[3] Much later, Ira Klein wrote a number of monographs explicating the interface between imperialism, ecology, mortality and epidemic cholera.[4] But such discourses are often mediated by considerations of political economy, and by factors such as polity, biology and ecology. What is generally overlooked or underscored in such engagements is the patients' perception: how did the victims perceive the epidemic? Did cholera epidemics play a part in the major political developments of nineteenth century India? At a different level, it may be asked what attempts were made by the people to reconcile the older notions of body humors and environmental miasmas with the new language of microbes and germs.[5] The present chapter would therefore arguably seek to explore how the proliferation of epidemic cholera and the consequent Western medical intervention in India were constructed in popular imagination.

ANTIQUITY AND INCIDENCE

According to popular belief, cholera is a disease which up to the beginning of the nineteenth century had never been seen before in

India. When Lord Amherst, the successor of Lord Hastings, reached India, he found that cholera was raging through the country. This fearful epidemic commenced its ravages at the beginning of the Maratha War in 1818, when it drew, for the first time, interested attention of European medical men.

Historically, cholera seems to have prevailed in South Asia from a very early period. It is mentioned in ancient Sanskrit texts under the names of *Sitanga* or *Vishuchi*.[6] The Hindustani name was *murree* or deadly disease, a word evidently derived from the same root as the Latin *mori* or English *murrain*. Cholera in many parts of India was called *murree* or *jurreemurree*, that is, the sudden pestilence, or *mahamurree*, the great pestilence.[7] The Sanskrit word usually believed to denote cholera was *bisuchika*. Macpherson argued in 1872 that originally it generally meant a disturbance of the stomach and intestines. Dr. Hessler, a nineteenth century Ayurvedic commentator, argued that it was certainly Ileus or spasm of the intestines, while Dr. H.H. Wilson, a Sanskrit scholar of the same period, translated it as spasmodic cholera.[8] Dr. Martin Haug thought that there were in Sanskrit different names distinctly descriptive of different stages of the disease: 1. *visuchika*, vomiting and purging; 2. *alasika*, cramps; 3. *vilambika*, collapse; 4. *dandasalika*, rigidity. In southern India, the spasmodic form of *visuchika* had been named *Sitangasonipat*.[9] With respect to treatment, Dr. Wise tells us that besides commencing with an emetic, and the application of the actual cautery to the argin, *Susruta* recommended a compound of myrobalan, orrisroot, assafoetida, the seeds of Wrightia anti-dysenterica, red garlic, rock salt, and *atees*, in equal parts. These were reduced to a powder, and then mixed with warm water for use.[10] Charaka, a later authority, prescribed an addition of opium and black pepper to the mixture. This recipe was said to have cured cholera even at the stage when the eyes were sunk, the pulse was imperceptible, and the extremities were cold.[11]

From 1761 to 1781, occasional outbreaks of cholera had taken place in various parts of India, but received scant attention. Epidemics were rare. It has been argued that cholera was endemic on the Lower Ganga River.[12] Pilgrims used to contract the disease at festival times and carry it back to the places of their return.

The first cholera epidemic seems to have broken out in the month of August 1817 at Kumbh Mela on the upper Ganga River in the town of Jessore, some 60 miles north-east of Calcutta.[13] The district of Jessore abounded in marshes, and was irrigated profusely by small streams and canals, which, when, stagnant, affected the surrounding atmosphere. Fevers and other disorders produced or promoted by unwholesome air were prevalent, especially during heavy rains or the inundations of the Ganga.[14] Nevertheless, epidemic cholera raged with destructive violence from 19 August, and so little was the nature of the epidemic known on its first appearance that there was extreme consternation. The civil courts of the district were shut down, and a temporary cessation of business of every description ensured.[15] People took flight. 'Such was the energy of the disease in this its first onset and so fatally destructive was it of human life, ' writes Jameson, 'that in this district alone, it is reported to have within the space of a few weeks, cut off more than six thousand of the inhabitants'.[16]

From Jessore the epidemic advanced up the river to Calcutta, where, after desolating the Black Town or the 'native' suburb, it spread through the major cities of Bengal, only sparing the elevated regions of Oudh and Rohilkhand. Throughout the month of August it devastated Dinajpur, Chittagong, Rajshahi, Bhagalpur, Mungher, Nattore, and Sylhet. On 15 September it was at Balasore, Purnea, and Cuttack; on 17 September it was at Buxar; on 18 at Chhapra and Ghazipur; on 5 November Mirzapur and Bundelkhand, and on 8 in the centre division of the grand army, encamped on the bank of the Sind.[17]

A detachment from the lower provinces introduced it into the army under Lord Hastings, then encamped on the banks of the Sind. The site was by no means salubrious, and did not afford a supply of good water. 'By the middle of November 1817', writes George Green Spilsbury, Bengal Medical Service, 'the cholera moribus seized the Centre division of the Army commanded by the Marquis in person. 10,000 fell victim, principally camp followers and about 30,000 deserted.'[18] Thousands succumbed to the attacks of 'the invisible foe', whose footsteps seemed 'shrouded in mystery', and therefore occasioned a more widely extended panic. Europeans and Indians alike fell under the scythe of the destroyer. The roads were covered with dying people and a 'melancholy silence pervaded the camp, interrupted only by the groan of expiring agony, or the passionate laments of despairing survivors'. In ten days nearly 9,000 human beings had perished.[19] Neither the European faculty nor the indigenous doctors, were able to discover the cause or the cure. 'The superstitious natives', writes a nineteenth century commentator, 'resorted to the expedient of making one more addition to the three hundred and thirty millions of their deities, and established rites to propitiate the malevolent goddess of the cholera.'[20]

Among the public and to a large extent the medical profession in the nineteenth century, the belief gained ground that the cause of this 'mysterious' disease was a specific germ, taken into the body under certain circumstances favourable to it. It was also believed that in certain persons who were not 'susceptible', the disease was of little consequence; but when it did take effect, it could be multiplied almost indefinitely all the resulting germs being as potent as those from which they had originated. These germs were supposedly given off by the excreta of those suffering from cholera; the disease could be spread only by the distribution of these germs either directly by human beings or indirectly through means of clothing or merchandise or other things that had become 'contaminated'.[21] Human beings, it was believed, were both the producers and the disseminators of cholera. The most common mode by which the germs were supposed to be distributed was the water supply. The excreta found their way into the water, and were thus widely and readily distributed among its takers. And from these beliefs the further conclusions had been formed that without human contact of one kind or another, it was impossible for cholera to spread. If human communication could be prevented, cholera would also be arrested. Therefore, all traffic of human beings or merchandise should be under a rigorous regulation of quarantine or medical inspection.[22]

It needs to be understood that, contrary to the popular belief, this was by no means a new affliction which distressed people in the district of Jessore in the year 1817. This disease could be distinctly traced in this country from the fifteenth century, and indeed, if records can be believed, it had been known from the earliest times. The description given in the early records of the symptoms—the suddenness of the attack, the vomiting and purging, followed rapidly by collapse and often within a few hours by death—are unmistakable.[23] In India of the nineteenth century cholera was exactly the same as it was at least five hundred years ago. There are records of a severe epidemic of it at Goa in 1543, where the Portuguese knew it under the name of *moryxy*. In the earliest European work on Indian medicine, published in Goa by Garcia d'Orta in 1563, the symptoms are fully described.[24] The disease was known by several other names, but it is important to note that attacks presenting all the symptoms which are now known as cholera had been recognized in India from the earliest times, and have prevailed with more or less severity up to the colonial days.[25]

It is only since the Sanitary Department was created that a systematic attempt was made to collect statistical data regarding this and other diseases in India. In 1868 the registration of death was introduced in the different provinces, and it was one of the main duties of the Sanitation Commissioners to improve this registration as the most valuable means not only of ascertaining the annual history of disease, but also to ascertain, from the death-rate, where sanitary reforms were most needed. One can see that the disease appeared year after year in nearly every part of the country, but that all the provinces had their years of epidemic prevalence as compared with years of marked abeyance. In Bengal, for example, the number of deaths fluctuated between 1,96,590 in 1876 and 39,643 in 1880.[26] It was an ailment, sometimes fearful and at times harmless. But during the colonial period cholera remained not just a disease per se. It carried many meanings—sometimes medical, social and cultural but more often than not political. Of all the issues involved in the epidemic, the question of contagion was hotly debated presumably because the government was anxious to protect its European population, both civil and military, from the 'Asiatic Cholera'.

DISCOURSE OF THE COLONIAL DOCTORS

The recurrent visitations of cholera in the first half of the nineteenth century called for urgency in the search for a cure. It prompted many medical men to explore the Ayurvedic sources with the help of *vaidyas*.[27] Texts like the *Chintamani*, an important indigenous medical source, were found to have reference to terms like *Sitanga* and *Vidhumar Vishuchi* which might be roughly approximated with symptoms of cholera.[28] One finds the reconciliation of the symptoms of *Sitanga* and *Vishuchi* in the term *Bisuchika* which became synonymous with cholera.[29] But practitioners of the dominant Western medicine dismissed the possibilities of finding a cure in the Ayurvedic texts. When Dr. Mahendra Lal Sarkar, the celebrated Allopath turned Homeopath, argued that camphor is one of the most important of the eleven to fourteen remedial agents for the first stage of treatment of cholera, his suggestion was brushed aside.[30]

Gradually, however, a belief gained ground among a few European medical men, particularly in Madras, that the disease was contagious. 'It has spread,' Mr. Scot believed, 'from the central part of Bengal to all adjacent countries; and though it may have appeared nearly simultaneously in many parts of Bengal, situated at considerable distances from each other, yet its progress beyond that tract has been uniform and progressive'.[31] Mr. Scot, therefore, concluded that either the disease had been propagated by infection or contagion, or that its progress was due to circumstances beyond human knowledge. The latter conclusion obviously left the question of the infectious or contagious nature of cholera undecided.

This precisely speaks much about the prevailing attitude of the government and the perception of their physicians. Throughout the nineteenth century the government was oscillating between positions taken by a section of contemporary medical men on the supposed contagious nature of epidemic on the one hand, and by others who posited the uncertainty of the contagion theory on the other. Cholera was considered a purely 'Asiatic' ailment; Europe was 'invaded' by this pandemic. Its origin in Oriental territories, its wayward yet determined progress across the globe, the novelty of its symptoms, and its rapidly fatal effects, invested it with a gloomy interest. To the Europeans, its parallel was to be found only in the chronicles of the Black Death or the Sweating Sickness of the middle ages.[32] British doctors of the nineteenth century used to believe that the ill fed and the impoverished were far more exposed to the assaults of this epidemic, as the digestive organs and the blood were principally affected.[33]

The specific cause of *Asiatic cholera* was generally sought in the atmosphere, but the precise condition in which it thrived had so far escaped detection. Dr. Prout, for instance, noticed a considerable increase in the weight of the air during the time that cholera prevailed in London, and this observation had been verified by similar discoveries made elsewhere. Both in Asia and Europe, fogs of an unusual kind, and of a very offensive odour, had been noticed. These atmospheric peculiarities and contaminations henceforth became the subject of more scientific scrutiny.[34]

The extent of its diffusion was one of the most striking characters of the epidemic. Cholera differed, it was noticed, from other epidemics in the irregularity with which it could pursue its march, as well as in its caprice in fixing upon isolated, and often very circumscribed localities. In some instances, however, the epidemic evinced a regular rate of progression; this was particularly observed shortly after it broke out in Bengal in the year 1817. In some districts of Bengal, it was observed, its course across the country was correctly calculated, at about one degree per month. Its subsequent itinerary did not preserve the same uniformity; on the contrary, the epidemic was frequently found to alight on locations more or less remote from its origin, leaving intervening areas untouched.[35]

Contemporary observers contended it did not appear that those countries where the system of quarantine was the most rigid had fared better. T.H. Starr, a nineteenth- century observer, reviewed the arguments and instances, and felt 'constrained to determine that its diffusion by contagion' was 'so improbable that we safely banish from our mind all the apprehension of danger from that source'.[36]

It was also observed that the rise and progress of cholera had in many places coincided with floods. This was particularly the case prior to its appearance in the delta of the Ganges in 1817 at Jessore, which for several months resembled a vast lake. The flood gradually subsided when, by the action of a vertical sun upon a vast overgrown swamp in the greatest profusion, the germs of this scourge seemed to have been endowed with powers of migration. The banks of rivers, often not navigable, had been in all latitudes the most favourable haunts of this epidemic. 'This fact', it was contended, 'harmonizes well with its origin from a swampy district, and leads us to the conclusion that a humid atmosphere is a good conductor of the disease.'[37]

Some doctors in the colonial administration believed that there had been an intimate connection between cholera and the consumption of unwholesome food.[38] The fact that the great majority of cases occurred during the night or very early in the morning, seemed to lend support to the theory. Such a belief was strongly rooted in the minds of Indian people; civil hospital assistants, as also the assistant surgeons, seemed to share the notion, and in reports submitted by them, foods received a prominent position in the etiology of the disease.[39]

Pilgrimage sites, especially Puri and Hardwar, were believed to be places highly choleric. During long marches, especially in the wet weather, pilgrims subsisted on parched rice, raw vegetables, unripe or overripe fruit, sweetmeats of every conceivable variety, containing a liberal admixture of fat, which they consumed in large amount. Other cases occurred which showed a period of starvation, followed by intake of bad food bought at the lowest possible prices and proving fatal, accompanied with symptoms whose diagnostic feature was supposed to be the presence of the 'comma bacillus'. So pilgrims seemed the prime suspects. In some villages in the districts of Puri there were no wells as such. Village tanks without exception were so foul that where well-water could be obtained for drinking purposes, tanks were reserved for bathing and washing clothes. Pilgrims, it was argued, were absolutely indifferent to the nature of water they drank when hot and thirsty, rushed for the nearest supply they could find, and considered it sufficient to cast aside the scum of filth that might have formed on the surface of the water contained in the tank or pool from which they drank.[40]

Precaution, in anticipation of the appearance of cholera had been a much talked about subject throughout the nineteenth century.[41] The outbreak of cholera was often so sudden and virulent that, it was argued, all precautionary measures must be taken beforehand. If cholera was reported in any town, the municipal sanitary superintendent should at once, it was insisted, get the people to clean up the town thoroughly. But if cholera had actually broken out, filth must not be disturbed or dug up. The best thing to do under these circumstances was to cover it up with fresh earth for a depth of six inches. The sanitary superintendent should advise the townsmen to have their houses immediately 'lime washed inside with hot lime wash'. Each house-holder was expected to burn inside each room in his house six ounces of sulphur in a pipkin.[42] The sanitary superintendent should always obtain from the dispensary and keep in a stock a sufficient quantity of medicine for diarrhoea as well as cholera. He should warn the townsmen that it was especially necessary that they should apply to him 'at once at any hour by day or by night' for medicine if any member of their family was attacked with looseness of the bowels.

Measures to be adopted on the appearance of cholera were meticulously enumerated. The municipal sanitary superintendent was advised to warn the people to abstain from drinking spirits. This was believed to be certain to promote the disease. It was also suggested that a thatched shed should be put up outside the town for the treatment of any wayfarers who might succumb to cholera. They should not be allowed to enter the town but the municipal superintendent should arrange that such persons were provided with suitable food, water, and medicine.

Sanitary specialists believed that those localities invaded by cholera must be looked on as dangerous. If possible, it was contended, the house should be immediately abandoned. It would be a measure of safety to move the patient away to a shed or open place outside the town, far from the place where he fell ill.[43] If possible, the patient should lie, on a cot and not on the ground. He should be given medicine according to the directions, but if the surface of the body was cold and the lips were blue, if the eyes appeared to be glazed, it was no good giving him medicine, but he should be well shampooed and the surface of the body be rubbed down with the palms of the hands. He should, if he could take it, be given hot mutton broth at frequent intervals. 'If his caste would not allow him to eat this', he should be given hot arrowroot conjee.[44]

The particular poison of cholera was supposed to be contained in the stools passed by the patient. It was therefore considered necessary that the stools be received into an old chatty or other receptacle and not spilt on the ground or bedding. But as soon as they were passed, they should be at once taken outside the town. A little straw and dammar should be placed over them, and they should be burnt, and the ashes should be buried and the same action should also be taken regarding vomited matter. If a patient died, the body should be at once burnt or buried in the clothes in which he had died; they should not be taken off the body.[45] The meticulous details with which the ailment was supposed to be handled indicate the nervous concern of the physicians working under the colonial government. Health of the troops was certainly a major component of their concern since cholera was taking heavy tolls of lives in the cantonment.

GOVERNMENT AMBIVALENCE

In view of the wide spread of cholera in the nineteenth century, the Government of India had ordered that quarantine should be enforced and continued as far as cantonments were concerned.[46] During the early years of the epidemic, the sick were in many places forcibly removed from their homes and placed under treatment in cholera hospitals, the object being the limitation of the spread of infection and 'the more scientific treatment of the disease'. But Dr. A.C.C. De Renzy, Sanitary Commissioner of Punjab, argued that the civil authorities should limit their efforts to hospital accommodation and medical treatment for those homeless and friendless poor, willing to avail of such accommodation. But on no account, he warned, should officials resort to any compulsion to remove the poor from their homes for treatment.[47] These measures he believed were mischievous in the extreme, for they led people to conceal the existence of the disease.

The question of contagion was hotly debated. G.S. Beatson, Inspector-General of Hospitals, Her Majesty's British Forces in India, argued in June 1873 that cholera in the ordinary sense might be considered to be non-contagious, but certainly 'it is contagious in the usual acceptance of the term in a very limited degree'.[48] W. Stewart, E. Goodeve and E.D. Dickson, the members of the British Cholera Commission concluded in 1866 that 'cholera is communicable from the diseased to the healthy, (a) by persons in the state of developed cholera, (b) by persons suffering from choleric diarrhoea who can move about and who are apparently in health for some days during the progress of the disease'.[49] Nevertheless, they concluded there was no reason to suppose that cholera was communicable by touch.

To the idea of contagion, many authorities in India had been strenuously opposed, but the belief in the communicability of cholera was gaining grounds. Major G.B. Malleson, Sanitary Commissioner for Bengal, argued in 1867 that in dealing with the question of practical measures for checking the disease, 'it is necessary that this communicability should be assumed as an established fact'.[50]

In India, special measures, founded on the transmissibility of the disease, were not taken into serious consideration until after the submission of the report of Mr. Strachey, President of the Committee of Enquiry into the Epidemic of 1861 in the North-Western Provinces. It was not until the late 1860s that any serious thought was given to the danger arising from pilgrimages even though the danger had been pointed out earlier by Dr. Greaves and others.[51] In the annual report of Bombay for 1863, Dr. Haines showed in details the probable influence of pilgrimages on the town of Bombay.[52] The Principal Inspector General of the Medical Department at Madras, in his report on the mortality at Madras in 1864 again attributed to individuals returning from the festivals at Conjeveram, Tirupati, and Trivallore, to the town of Madras, in the same way that he had already, in 1860, attributed the propagation of cholera through the Central Provinces to the pilgrims returning from the temples of Mahadeo. Dr. Leith, President of the Sanitary Commission of Bombay, tells us in 1866 that in consequence of measures taken for the prevention of the ravages of cholera among the pilgrims, a remarkable result was obtained; there were ninety-four places of pilgrimage, and though cholera raged in the presidency, it broke out in only *two* of these places: Jeypoorie, where 5,000 devotees were assembled, and at Sumgum where the crowd amounted to 50,000.[53]

The measures taken for the prevention of the dangers arising from pilgrimages were of two sorts; first, measures to prevent the development of cholera at the places of pilgrimages during the stay of the pilgrims; second, measures to prevent the propagation of the disease by pilgrims on their way back. In Bombay, a huge number of camps was established for the pilgrims so as to prevent crowding in the towns; there were hospitals for pilgrims and a system of cleanliness; latrines were constructed, which, in some places, consisted merely of trenches dug in the earth to leeward, at a convenient distance from the camps, and filled with earth after being used. In regard to the return of the pilgrims, the Government of Bombay applied Section 271 of the Penal Code, and caused the entrance of the pilgrims into the towns and military stations to be watched. It required, before permission to enter was given, that proof should be afforded that people were not suffering from diarrhoea or other indications of cholera, and that they had had no communication for 48 hours with any persons showing such symptoms. In the absence of such proof, the pilgrims were retained under observation for 48 hours. In the contrary event, the sick were separated from the healthy, and the latter had to recommence the quarantine of 48 hours. In the application of these measures, arrangements were made for supplies to the pilgrims of provisions, shelter, and even medical attendance. If there were no means of supplying them with ordinary tents, tents were made in the indigenous fashion.[54]

Nevertheless, the International Sanitary Conference held in August 1866 advised upon other precautions, which were not attended to in Indian situations. Latrines had to be constructed to the leeward of the encampment; these might consist simply of trenches of a foot and a half in depth by as many in breadth; these latrines to be the only places resorted to by the pilgrims and the pilgrims were required, immediately after having made use of them, to cover their excrement with earth. The sick had to be separated from the other pilgrims, and disinfectants applied to their excreta, such as the solution of permanganate of potash, chloride of zinc, carbolic acid, or if these substance were not available, quicklime.[55]

The colonial rulers, it appears in retrospect, were in a real dilemma. They realized through hard experiences that the process of colonization should begin not just by battles with the people targeted, but also against the diseases they were subjected to. The informed apprehensions aired by officers about the quick spread of epidemic cholera could not possibly be allayed simply by medical intervention and the material resources at their command. They needed to take into account several other things, of which people's perception about the disease was perhaps the utmost one.

POPULAR CONSTRUCTIONS OF THE EPIDEMIC

To many Indians, cholera was an inalienable outcome of British military conquest achieved by gross violation of Indian religious sensibilities. Villagers in Bundelkhand, for instance, believed that the epidemic had begun because despite a Brahmin's remonstrance, cattle were slaughtered to feed British soldiers, encamped in a grove, sacred to Hurdoul Lal, son of a former local raja.[56] In people's perception, the move amounted to a violation of Hindu tradition on two counts: first, it vitiated the sanctity of a sacred grove, and second, it outraged the Hindu prohibitions on the killing of cow and eating of beef. The epidemic was, therefore, interpreted as a sign of disorder caused by divine displeasure. To remedy the situation, Hurdoul Lal was worshipped and propitiated over a large area of northern India whenever cholera broke out.[57]

Perception of this kind spread far and wide. Villagers at Cunnatore in Nellore district of Madras, for instance, sought to relate the epidemic to the coming of the 'low-caste soldiers' and their encampment near a sacred tank.[58] The inhabitants of Ongole in the same district too attributed epidemic cholera to the maltreatment by the British of the raja of Gumsur. Taken captive by the British, the raja failed to make his usual puja to the goddess Mahalakshmi. The goddess retaliated by inflicting the disease on both the British troops and the people. Moreau de Jonnes reported that the Hindus attributed the 1817-21 epidemics in a similar way to the resentment of the deity 'Yagatha Ummah' against British rule in India.[59] Thus one can see that the prevalent popular belief was that the British were responsible for spreading the disease through the direct violation of Hindu taboos as also through their military intervention on the Hindu cosmos.

Epidemic cholera created such consternation that several narratives of disease deities were floated. David Arnold argues that although it became clear during the early nineteenth century that cholera had long been present in India, the disease had not, unlike smallpox, been extensively ritualized. It is difficult to accept this position. In lower Bengal, for instance, we find that people had for a long time past worshipped the goddess of cholera as the Ola Bibi. Macnamara had taken considerable pains to make out the history of the worship, and had on several occasions not only visited the temple of the Ola Bibi in Calcutta but had very good photographs of the building and of the idol it contained.[60] It appears, according to tradition, that at an early period, a woman, while wandering about in the woods met a large stone, the symbol of the goddess of cholera. The worship of the deity through this stone was, according to the prevailing ideas of the Hindus, the only means of preservation from the influence of this terrible disease. The fame of the goddess spread and people flocked from all parts of the country to her shrine in Calcutta.

As was usual in such cases, the idol became the property of a family and a source of income. Originally, the idol Macnamara referred to, was kept under a bamboo shed; but early in the eighteenth century, probably the year 1720, an English merchant, to please his 'native friends', built a temple to the goddess.[61] Of the rites performed at the shrine, we know that besides presenting offerings, the votaries of the goddess fasted in the morning, and at about 2 o'clock in the afternoon, dined upon crushed rice and *dahee,* eating nothing after that until the next day. Every Tuesday and Saturday, Macnamara relates, some three or four hundred women used to worship after this fashion, and returned to their respective homes in the evening. The pilgrimage was especially common from April to June, the cholera season.

With the passage of time, the temple location became inconvenient, and Mr. Duncan donated Rs. 6,000 for the erection of a structure built probably in 1750; near it, was the tomb of Mr. Duncan's Mohammedan wife and child. The old crude stone was transferred to the new temple, and a somewhat elaborate idol was constructed; it represented in the centre a carcass with a preying vulture; on the back of the latter, the goddess was represented with four hands and in a sitting posture; on the right was Manasa, the goddess of serpents; next to her, is Shiva, the destroying principle; next comes a female in a supplicant posture, and a male afflicted with the disease.

The female is supposed to be praying to Shiva for the recovery of her husband. On the left of the goddess, are the idols of Sitala, the goddess of smallpox, and of Shasthee, the goddess presiding over infants and children. The temple continued to be the property of the family that originally possessed it, and it was producing an income of Rs. 300 or 400 a year.

From this curious narrative of Macnamara, we are entitled to infer, although cholera might not have been so prevalent in India in the commencement of the eighteenth, as it was in the seventh century, nevertheless, for years past it had been a well-known disease in Bengal. It seems also certain that when the disease raged at times with violence, people found it urgently necessary to propitiate the deity. But Dr. Sambhu Nath De had a different story to tell us.[62] He argued that prior to 1817, temples of Ola Bibi was absolutely non-existent in Calcutta. The word 'Ola', he argued, had originated from Jessore which means that the disease and its name had had a migratory past. The truth, perhaps, seems to lie somewhere in-between. It appears that the worship of Ola Bibi in a regular temple structure might have been an event not earlier than the nineteenth century, but her worship or propitation might have been a much older tradition.

Instances of popular belief in the power of a deity to cause harm or control this disease are coming in abundance. One expression of this belief was the appearance of a young woman dressed as the goddess of the disease and worshipped accordingly. Another was the appearance of a woman claiming to be medium possessed by the spirit or deity responsible for the disease. Both acted as prospective mediators between the villagers and the gods, voicing the grievances of the goddess and giving an immediate focus to local anxieties and alarm.[63] Instances of such mediums or *devis* are prolific in anthropological literature, and they seem to characterize the popular approach to an epidemic. But the cholera epidemic of the nineteenth century exhibited a degree of social unease and tension almost unmatched by any other epidemic in India.

In January 1818 the Reverend J. Keith reported the appearance of 'an actual avatar or incarnation of Ola Beebee' at Salkia on the outskirts of Calcutta.[64] She sat out there for two days 'in all the state of a Hindoo goddess' attended by a young Brahmin woman as her 'priestess'. Keith tells us that she competed with the goddess Kali for the attention and offerings of worshippers alarmed by the epidemic; she succeeded in scaring the people. In the same year in the Bombay Presidency too, a woman declaring herself to be an *avatar* of the fiend of pestilence was seen at the cantonment of Sirur near Pune. Kennedy tells us that she entered the bazaar almost naked.

Her disheveled hair, her whole body, and her scanty apparel were daubed and clotted with the dingy red and ochry yellow powders of the Hindoo burial ceremonies. She was frantic with mania, real or assumed, or maddened by intoxication, partly mental, partly from excitement from drugs. In one hand she held a drawn sword, in the other an earthen vessel containing fire (the one probably a symbol of destruction, the other of the funeral pile). Before her proceeded a gang of musicians, pouring forth their discords from every harsh and clattering instrument of music appropriate to their religious processions. Behind her followed a long line of empty carts; no driver whom she encountered on the road daring to disobey her command to follow in her train. Thus accoutred and accompanied her frenzy seemed beyond all human control; and as she bounded along, she denounced certain destruction to all who did not immediately acknowledge her divinity; and pointing to the empty carts, which followed, proclaimed that they were brought to convey away the corpses of those who rashly persisted in infidelity. No ridicule, no jest, awaited this frantic visitant, but deep distress and general consternation.[65]

Babington has narrated still another event during the same epidemic. It involved Christian Koli fishermen in the village of Chendnee in northern Konkan.[66] When cholera struck in 1818, many of the Kolis either died or fell sick. They appealed to the government for help, but government medicine too proved ineffective. They asked their Catholic priest to provide a remedy for the evil, and when he too failed, they resorted to their traditional treatment, called *khel*. The villagers formed a circle round a number of people, mainly women, whose groans and violent gestures seemed to indicate that they were under a 'supernatural influence'. They were sprinkled with water and coloured earth, and urged to dance frantically at the sound of 'native music'. The women in trance became mediums of several spiritual visitors. The mediums pointed to certain villagers who had supposedly caused the epidemic; they were severely punished. Several Kolis, stricken with cholera, were said to have been cured merely by sprinkling earth and water over their bodies.

Instances of such episodes can be multiplied, and Babington, Kennedy, Keith, and other colonial magistrates, doctors, and missionaries had indeed done so.[67] But they did so, as David Arnold has rightly argued, to demonstrate the 'primitive, superstitious, and irrational ways of Indians and to commend the superiority of their own medical practices or religious beliefs'.[68] But this position was counter-productive. Opinion about the ailment's nature, mode of transmission, the effectiveness of sanitary measures and even the efficacy of Haffkine's anti-cholera serum were so divided that the government finally reverted to a safer position of non-intervention. It was unwilling to damage its commercial and political imperatives.

Back at Bengal, we have had varying tales of wails and woes, fear and fright, coming as they did from the epidemic in the countryside, reflecting much of its horror in the contemporary Bengali literature. And, the novels of Sarat Chandra Chattopadhyay fit the bill the best. Sarat Chandra provides us with an altogether different archive of knowledge, not just on cholera, but on the process how a society functions when a pestilence strikes terror among its inhabitants. Cholera, better known in Bengal as *Bisuchika*, figures in *Pandit Mashay* (1914) with such a force that it becomes a character itself, pulling events in and out, and finally forcing the novel's inevitable end. Almost all the characters in the novel come into focus when cholera breaks out in the village in an epidemic form, killing people young and old within a very short span of time. Villages in Bengal, often portrayed in literature as idyllic, eking out a sheltered life in complete isolation, free from greed and devilry, are represented by Sarat Chandra in a different hue. Cholera sharpened caste animosities, raked up inter-personal rivalry, and broke down traditional social relations beyond repair.

The epidemic breaks out perhaps owing to contaminated drinking water drawn from the only tank in the neighbourhood, which is owned and maintained by Brindaban, the *panditmoshay* (village school master). One of his neighbours, a Brahmin by caste, dies of cholera. Brindaban notices that the clothes of the deceased are being washed clean in his tank, the water of which is shared by many. He objects on the ground of hygiene and public health. But villagers used to believe that according to Hindu scriptures water of a tank dug ceremoniously can never be polluted. Brindaban argues, 'this is the only pond in the village; I shall not let it get spoiled in this trying time.' Tarini Mukherjee, the contending neighbour, requests Ghoshal to intervene. Ghoshal, whom the villagers believe to be the sole guardian of Hindu religious practices bursts out in anger, 'this is your wrong insistence, Brindaban; according to *shastras* pond water can never be polluted or made unholy! You may have read a few pages in English, but how can you disbelieve our *shastras*?'[69] The Brahmin makes it an issue, makes common cause with other village notables, and holding out his sacred thread curses him: his whole generation will be wiped out through cholera! Eventually, Brindaban's mother, an old pious lady, dies of cholera, and a few days later his only son also falls to this dreaded disease.

Cholera takes its toll of human life awfully fast, and its fury is so dramatic and spectacular that it puts to test the age-old superstitions, customs and regulatory social practices in Bengal. The term 'Cholera' or *ola-utha* was invoked to curse a belligerent neighbour: 'to you and to family let cholera come, and you and your family get ruined!' It was supposed to be the most unkindest curse of all.

In Bengal, more often than not, cholera drove people to seek remedies in divine therapy. Tarasankar Bandyopadhayay's *Arogya Niketan* depicts how outbreak of cholera spread panic in the countryside.[70] Wild rumours were afloat far and wide: the cholera deity or Olai Chandi can actually be seen; she roams the village in the evening. It is a girl, dragging her lean and thin body with fiery eyes and disheveled hair; she wears a damp and ragged cloth; tucked under her arms a tattered mat, she trudges along the margin of a muddy village road. She prefers to remain incognito, but whoever notices her first becomes her victim. After taking its first toll, the disease travels fast from house to house, and from village to village. At midnight, street-dogs bark, wailing for long, indicating that they have noticed the unfamiliar girl roaming. The barking becomes louder.[71] To propitiate the deity people resort to community worship of Ola Bibi or Olai Chandi with all the rites and rituals observed carefully. In the early evening, they take out a religious procession in the village lanes, singing out songs praising the virtue of the goddess Ola Bibi or Olai Chandi.[72]

In Bengal, Olai Chandi was generally worshipped on Tuesdays and Saturdays throughout the whole cholera season. Though she is worshipped all over Bengal, her worship seems to be much in vogue in Nadia and 24-Parganas where she is worshipped along with Raksha Kali. In Tamil Nadu, cholera deity is known as Mariyamma or Ankamma, while in Orissa she is worshipped as Jogini Devi. Dipesh Chakrabarty argues that such deities as Ola Bibi or Mariamma are invoked to protect a small geographical area, and therefore the rituals associated with their worship help convey a sense of 'locality'.[73]

CONCLUSION

Epidemic cholera in nineteenth century India marked the widening gulf between indigenous and Western medicine.[74] While many Indians sought help in religious observances rather than in medical concoctions, *vaidyas* and *hakims* provided their assistance with medicines consistent with their humoral pathologies. The *vaidyas* prescribed medicines composed of black pepper, borax, aniseed, ginger, and cloves. Some resorted to opium to neutralize the pain and relax the body. British doctors took careful note of these indigenous therapies hoping to profit from the treatment of a disease unresponsive to their own therapeutics. Despite their cultural reservations, they shamelessly borrowed from the *vaidyas* and *hakims* or administered almost similar drugs. Dr. Jameson in Bengal, for instance, in 1820 discussed Indian doctors' treatment for cholera at length and approved of the use of opium and calomel. By the 1850s a 'cholera pill' was made from opium and black pepper. But even though they openly borrowed from indigenous practice, they remained disposed to believe in the superiority of their own therapeutics. However, these remedies, whether prescribed by Indian or European physicians, were not always received well by the patients, and often failed to save the sick. Religious rituals alone seemed appropriate.

Nevertheless, cholera did not cease to carry meanings of its own for the nineteenth century Indians. It became much more value-loaded than ever before. It turned out to be a 'highly political disease'. That cholera proved to be 'exceptionally troublesome disease, unresponsive or resistant to most of the favoured therapies of the time' made it politically and culturally all the more different from the prevailing ailments.[75] It remained a memory for both the colonized and the colonizers. To the colonizers, memory of cholera catapulted a political reading that enforced state intervention to the minimum. To the colonized, cholera bequeathed a comprehensive memory of physical frailty and political vulnerability of colonial rule. The skill of the physicians, the advice of the hygienists, the resources of the public health engineers, and the capabilities of the rulers were put to test by the widespread epidemic.

NOTES

1. Asa Briggs, 'Cholera and Society in the Nineteenth Century', *Past and Present*, no. 19, 1961, pp. 76-96.
2. C.E. Rosenberg, *The Cholera Years: The United States in 1832, 1849 and 1866*, Chicago: University of Chicago Press, 1962; R.E. McGrew, *Russia and the Cholera, 1823-32*, Madison: University of Wisconsin Press, 1965; M. Durey, *The Return of the Plague: British Society and the Cholera, 1831-32*, Dublin: Gill and Macmillan, 1979; F. Delaporte, *Disease and Civilization: The Cholera in Paris, 1832*, Cambridge: MIT Press, 1986.
3. David Arnold, 'Cholera and Colonialism in British India', *Past and Present*, no. 113, 1986, pp. 118-51.
4. Ira Klein, 'Imperialism, Ecology and Disease: Cholera in India, 1850-1950', *The Indian Economic and Social History Review*, vol. 31, no. 4, 1994, pp. 491-518; 'Death in India, 1871-1921', *Journal of Asian Studies*, vol. XXXII, no. 4, pp. 639-59.
5. Deepak Kumar, 'Social History of Medicine: Some Issues and Concerns', in *Disease and Medicine in India: A Historical Overview*, Deepak Kumar, ed., New Delhi: Tulika, 2001, pp. xiv-xvii.
6. F. Hewitt, *The History of British Settlements in India*, London: no pub., 1855, pp. 272-4, cited in *Extracts from Records of Past Epidemics in India, 1912*, National Archives of India (hereafter NAI), New Delhi.
7. John K. Macpherson, *Annals of Cholera: From the Earliest Periods to the Year 1817*, London: Ranken and Drury, 1872, p. 6.
8. Ibid., p. 9.
9. Ibid.

10. T.A. Wise, *Commentary on the Hindu System of Medicine*, London: Forgotten Books, 1845, p. 330.
11. Macpherson, op. cit., p. 26.
12. J.N. Hays, *Epidemics and Pandemics: Their Impacts on Human History*, Santa Barbara, California: ABC-CLIO, 2005, p. 193.
13. Hewitt, op. cit.
14. Arabinda Samanta, *Malarial Fever in Colonial Bengal: Social History of an Epidemic*, Calcutta: Firma KLM, 2002.
15. Frederick A. Corbyn, *Treatise on the Epidemic Cholera, as it has Prevailed in India; Together with the Report of the Medical Officers, Made to the Medical Boards of the Presidencies of Bengal, Madras and Bombay, for the Purpose of Ascertaining a Successful Mode of Treating that Destructive Disease; And a Critical Examination of All the Works which have Hitherto appeared on the Subject*, Calcutta: Bengal Establishment, 1832. p. 5, T 11417, OIOC, BL, London.
16. Ibid.
17. Ibid., p. 6.
18. Letters dated 1811-24 of George Green Spilsbury (1785-1857), Bengal Medical Service, 1811-57, to members of his family in England. *MSS Eur. D. 909*, OIOC, BL, London.
19. Hewitt, op. cit., p. 274.
20. John Clark Marshman, *The History of India*, London: Longmans, Green, Reader and Dyer, 1867, pp. 329-30. OIOC, BL, London.
21. Cuningham, op. cit., p. 3.
22. Ibid.
23. Jadunath Mukhopadhyay, *Bisuchika Roger Chikitsa*, Chinsura: Chikitsa Prakash Press, 1872, VT/949, OIOC, BL, London.
24. Cuningham, op. cit, pp. 4-5.
25. Ibid., p. 5.
26. Ibid., p. 7.
27. Dhrub Kumar Singh, 'Choleric Times and Mahendra Lal Sarkar: The Quest of Homeopathy as "Cultivation of Science" in Nineteenth Century India', in *Medizin, Gesellschaft und Geschichte*, MedGG 24, 2005, p. 209.
28. William Scott, *Report on the Epidemic Cholera as it Has Appeared in the Territories Subject to the Presidency of Fort St. George, Drawn up by the Order of the Government under the Superintendent of Medical Board*, Madras, 1824.
29. Singh, op. cit., p. 209.
30. Ibid
31. Corbyn, op. cit., p. 71.
32. Thomas Henry Starr, *A Discourse on the Asiatic Cholera, and its Relations to*

some Other Epidemics Including General and Special Rules for its Prevention and Treatment, London, 1848, p. 9. OIOC, BL, London.

33. Ibid., p. 13.
34. Ibid.
35. Ibid., p. 18.
36. Ibid., pp. 22-5.
37. Ibid., p. 30.
38. Charles Banks, *Observations on Epidemics of Cholera in India, with Special Reference to their Immediate Connection with Pilgrimages*, Cuttack: no pub., 1896, OIOC, BL, London.
39. Ibid., p. 24.
40. Ibid., p. 27.
41. E.F. Barretto, *Instructions in Regard to Cholera Epidemic*, Agra: no pub., 1892, p. 7. tr. 735, OIOC, BL, London.
42. Ibid., p. 10.
43. Ibid., p. 12.
44. Ibid., p. 13.
45. Ibid., pp. 14-15.
46. Government of India (hereafter GOI), Home Department, Sanitary Branch, March 1874, NAI, New Delhi, Proceeding no. 14.
47. From A.C.C. De Renzy to T.H. Thornton, 5 June 1873, GOI, Home, Sanitary, March 1874, NAI, New Delhi.
48. From G.S. Beatson, to F.S. Roberts, Officiating Quarter Master General of the Bengal Army, GOI, Home, Sanitary, March 1874, NAI, New Delhi.
49. Conclusions of the British Cholera Commissioners, India Office, London 16 July 1866: Forwarded for information to the Govts of Bengal, NW Provinces and the Punjab, the Chief Commissioners of CP and British Burma, etc. Home Dept. Public, B Proceedings, August 1866, NAI, New Delhi.
50. Major G.B. Malleson, Sanitary Commissioner for Bengal, to the Secy to the GOI, Military Dept. (no. 312 dated Simla, 29 May 1867), Home, Public, August 1867, NAI, New Delhi.
51. *International Sanitary Conference: Report on the Hygienic Measures to be Adopted for Preservation against Asiatic Cholera with an Appendix on Disinfections as Applied to Cholera*, Calcutta (n.d.), NAI, New Delhi, p. 5.
52. *Death in Bombay During 1863*, vide *International Sanitary Conference*, op. cit.
53. Ibid., p. 6.
54. Ibid.
55. Ibid.
56. David Arnold, *Colonizing the Body: State Medicine and Epidemic Disease*

in Nineteenth-Century India, Berkeley, Los Angeles, London, University of California Press, 1993, p. 171.

57. W.H. Sleeman, *Rambles and Recollections of an Indian Official*, vol. 1, London: J. Hatchard and Son, 1844, p. 211.

58. William Scot, *Report of the Epidemic Cholera as It Has Appeared in the Territories Subject to the Presidency of Fort St George, Madras*, 1824, p. 237.

59. Arnold, *Colonizing the Body*, p. 171.

60. C.A. Macnamara, *History of Asiatic Cholera*, London: Macmillan, 1876, p. 34.

61. Ibid., p. 35.

62. S.N. De, *Cholera: Its Pathology and Pathogenesis*, Edinburgh and London: Oliver and Boyd, 1961, p. 27

63. Arnold, *Colonizing the Body*, p. 172.

64. Reverend J. Keith, Calcutta, to London Missionary Society, 1 January 1818, London Missionary Society Archive, School of Oriental and African Studies, London, Cited in David Arnold, *Colonizing the Body*, p. 173.

65. R.H. Kennedy, *Notes on the Epidemic Cholera*, Calcutta: Baptist Mission Press, 1827, pp. ix-x.

66. Arnold, *Colonizing the Body*, pp. 173-4.

67. Ibid.

68. Ibid.

69. Sarat Chandra Chattopadhyay, 'Pandit Masai', *Sarat Rachana Samagra*, Kolkata: Basak Book Store, 2008, p. 54.

70. Tarasankar Bandyopadhyay, *Arogya Niketan*, Kolkata: Prakash Bhavan, 1359 (BS), p. 145.

71. *Arogya Niketan*, p. 146.

72. Ibid.

73. Dipesh Chakrabarty, 'Shareer, Samaj o Rashtra: Oupanibeshik Bharatey Mahamari o Janasanskriti', in Gautam Bhadra and Partha Chattopadhyay eds., *Nimnabarger Itihaas*, Kolkata: Ananda Publishers, 2001, pp. 169-71.

74. The differing methods of treatment by the European and the indigenous doctors has been addressed by Projit Bihari Mukharji in *Nationalizing the Body*, Chapter V, pp. 179-212.

75. Arnold, *Colonizing the Body*, pp. 198-9.

CHAPTER 3

Smallpox

Literature on smallpox epidemics in colonial India and particularly in Bengal is coming up in abundance. Much has been written on the subject from various standpoints. Some scholars have emphasized the killer disease, keeping in mind the medical encounter of the physicians as also the medical research undertaken by the colonial masters,[1] while others have examined the phenomenon in a socio-historical perspective.[2] Some again have looked into the epidemic in terms of the emergence of colonial medicine in India.[3] Still others have engaged the scourge to address the question of vaccination in an uninitiated society.[4]

Much of the historical engagement with epidemic smallpox in colonial Bengal/India is generally caught between the rhetoric of two opposing sites of intervention: the inverted polarity of inoculation-vaccination syndromes. From Ralph W. Nicholas[5] and David Arnold[6] to Michael Worboys, Mark Harrison and Sanjoy Bhattacharya,[7] the discourse swings between the 'poetics' of worship and rituals associated with Sitala, the reigning deity of smallpox, and a 'technique' of inoculation. Though wisdom, individual or collective, and perception, secular or religious, are verily present in their discourse, these are eventually marginalized in the entire exercise. Despite the 'native' perception of the failure of vaccination, it has been argued, medical establishments continued to sustain the rhetoric of vaccination, presumably because state sponsored vaccination sought to redefine the contract between patient and practitioner.[8] Under colonial dispensation, a vaccinator ceased to be a practitioner altogether. In popular perception, he was just a marker of bodies and an extractor of lymph. To them, the pock mark was a new sacrament, 'an external visible symbol of an internalized state', indicating that the person had not only 'taken' the disease and was protected but was also the carrier of an externally visible sign of a contract he had entered into vis-à-vis the state. All of them generally shared a common concern that smallpox in colonial India assumed a political meaning. Missing in the recent discourse is the patient's perception. The moot question is how did the people perceive the killer disease? Some have looked into the magico-religious rites and rituals in connection with epidemic smallpox in Bengal.[9] How can one trace the interface between the great tradition and little tradition in terms of folk medicine and ethno-history?[10] Larger than this is the crucial question, how did the sick evaluate a doctor? Literature on smallpox is indeed prolific, but we need to look at it from the patients' perceptions, their collective imagination, and their cultural tradition.

ANTIQUITY AND INCIDENCE

Smallpox had all along been a general disease in India. People did die from it, and even though many recovered, they were disfigured by it, or even blinded by it. In order to diminish the disease, inoculation was resorted to in parts of India. Matter was taken from a person with smallpox and inserted under the skin of a healthy person. People who were thus inoculated did get smallpox but in a much milder form than otherwise. But unfortunately inoculation/variolation was not without danger. Sometimes a very severe attack followed and a person could die.[11] Both inoculation and modern vaccination are forms of immunization. Inoculation is a method of introducing a live organism in the body in a controlled way so as to minimize the severity of infection and induce immunity in future. Vaccination is a process of introducing a weakened form of the pathogen to a healthy body so that the immune response is triggered to fight the actual pathogen in future.

The incidence of smallpox in India was great and uniform from year to year ever since its presumably first historically recorded virulence in Assam in 1574.[12] Stavorinus, a Dutch naval commander in Chinsura, recorded an incidence of severe type of smallpox in Bengal in 1770, which he argued, spread among persons of all ages and killed people in great numbers.[13]Holwell wrote in 1767 about the prevalence of smallpox in Bengal where every seventh year, without respite, smallpox raged during the months of March, April, and May, until the annual returning rains, about the middle of June, put a stop to its fury.[14]

But references in early medical compilations to a disease called *masurika,* a kind of pulse similar to the kind of skin eruptions during the disease, suggest that it was an ancient affliction. *Basanta roga* was also known as *paproga, shitalika, shitala, gunri,* and *guli.*[15] Even *masurika,* we are told, was of nine varieties: *bataja, pittaja, slesmaja, tridosaja, batapittaja, bataslesmaja, pittaslesmaja, raktaja,* and *charmaja.*[16] The term 'small pokkes', meaning pockets and bags, as distinguished from great pox 'syphillis', was introduced in the country in the fifteenth century by the Portuguese. Due to its prevalence during spring, the ailment was also known in eastern India as the *Basanta Roga* or the spring disease.[17] Festivities during *Basanta* helped transmit the disease in all directions. Moreover, the political disintegration of the Mughals facilitating the British acquisition of power, the recurring Maratha raids resulting in displacement of population, the severe drought of 1769 and the consequent famine of 1770 would also have favoured its dissemination.[18] Smallpox was thus a scourge, responsible for more victims than all other diseases combined, outstripping even cholera and plague in its tenacity and malignancy.[19]

It is difficult to determine the exact figures for smallpox mortality in early nineteenth-century rural India due to the paucity of census data. We have, of course, some official statistics of smallpox ravages for Calcutta. The Report of the Smallpox Commissioners, Bengal, stated in 1850 that within eighteen years smallpox appeared to have visited Calcutta in an epidemic form no less than four times, each for about twelve to sixteen months. During the intervening periods, the disease seemed to have entirely disappeared.[20] Its sudden reappearance so often in a populous and developing city, at a particular season, and when the surrounding villages were free from the disease, might lead to the inference that the occurrence had not been accidental or of natural causes. But whatever might have been the reason, the fact remains that the mortality was enormous. The epidemic in question, the Commissioners argued, had very far exceeded all its predecessors both in fatality and in duration.[21] The 1832-3 epidemic carried off 2,814 people in sixteen months in Calcutta; in 1837-8, it killed 1,548; in 1843-4, 2,949; but in 1849-50 it swept away no less than 6,100 people.[22] Smallpox was the most destructive of Indian diseases.[23]

In his Annual Report on Vaccination to the Medical Board for 1838, Dr. Stewart computed that

taking the census of Calcutta made in 1837 to be correct and the average of six years as affording a fair estimate of the mortality, the annual mortality of Hindoos by smallpox is .295 per cent, or one in 399; that of Mohammedans is only .128 per cent, or one in 782. Out of 100 deaths of Hindoos, five and a half are caused by smallpox; in 100 deaths among Mohammedans, the number caused by smallpox is six.[24]

GOVERNMENT INTERVENTION AND VACCINATION

The government sought, in time, to replace inoculation by Western vaccination. Much before the statutory prohibition of inoculation in Calcutta and its suburbs in 1865-6, *ingrajitika* or vaccination was introduced in Bombay. Dr. Edward Jenner who propagated vaccination in 1798 was keen to send it to India and volunteered a shipment of the vaccine with an offer of thousand pounds. The Governor of Bombay, meanwhile, had appealed to Lord Elgin, and received in 1801 a quilt containing lymph.[25]

But the history of vaccination in Bengal presents a shoddy picture, resulting from the want of principles of organization and of perseverance in the execution of any general scheme. It exhibits at one time a struggle between efficiency and economy, at another a vain attempt at compromise between the alleged prejudices of ignorance or superstition, and the views of a seemingly enlightened benevolence. Sanjoy Bhattacharya, Mark Harrison, and Michael Worboys together argue that the development of smallpox controls and public health policies between 1890 and 1940 mirrored the fractured nature of the colonial Indian administrative structures.[26] They argue that conflicts arose frequently between British bureaucrats, and within government departments, such that even when adequate funds were available, vaccination was occasionally impeded by the competing interests of various government officials. The promotion of vaccination, it is true, had never ceased to engage the attention of successive Governor-General, each new Medical Board, and all the Superintending Surgeons, but until the closing years of the nineteenth century it had made no corresponding progress in public opinion, and its practical benefits were still almost entirely confined to the Europeans.

Let us see what transpired between the givers and takers of vaccination. The vaccinators, it can be argued, were indeed not government servants and as such were not recognized by the people. They were looked down upon. They depended for their living on the fees they received from the people. Unfortunately they could not realize those fees in many cases, as there was no special law making fees compulsory.[27] They had, therefore, to depend on the good graces of the village *panchayats* or the police in the matter. But the people had come to know that the only course left to the vaccinator to realize his fees was to institute a civil suit, which he would not care to do for the worry and expense of it. And they thus easily evaded payment. The result was that many trained, good vaccinators threw up their appointment in disgust, and took to some other occupation for a living. It was also difficult to replace them as there was no competition for the work. Those who stuck to the occupation worked in areas where fees could be easily obtained, leaving others to do so where there was opposition to vaccination and the people were reluctant to pay fees. So many places were left 'unprotected'.[28]

From Dr. Cameron's Report as also from the Records of the Medical Board, we learn that in 1828 there were as many as thirty vaccine stations maintained by the Government in the Bengal Presidency at an average expense each of Rs.260 per month. Dr. Stewart suggested that the Board should concentrate its efforts on certain towns between Calcutta and Delhi in order to keep all parts of the country regularly supplied with fresh virus. The Board expressed its approbation, but reserved its comments on the proposed employment of six assistant surgeons entirely to vaccinate. It recommended instead that the expensive establishments then maintained at sixteen stations should be reduced to Rs. 120 each, and that eighteen new vaccine stations should be formed at an expense of Rs. 60 each, while Calcutta,

Dacca and Patna alone should be continued on the existing scale of allowances.[29]

Dr. T Smith, a senior member of the Medical Board wrote to the Governor of Bengal in the same year arguing that the salaries of the Civil Surgeons were insufficient to induce them to devote their mind to the profession, or even to prevent them from 'partly renouncing it by becoming a merchant'.[30] Dr. Stewart was mistaken, Smith apprehended, in attributing to natives generally a partiality for European remedies and practices. The rich and higher classes, Smith argued, unless 'enlightened by European education and habits', invariably preferred the prescriptions of their own *hakims* and *vaidyas* to those of European physicians. The indigent, at least equally prejudiced and ignorant, would express the same preferences 'if they had the means of indulging them', i.e. if they could pay their own practitioners. But under pressure, and not choice, they would accept offers of gratuitous treatment and attendance from the Europeans. As for the dislike which the people of Bengal usually showed for the necessary discipline of Government Hospitals, Dr. Smith considered dispensaries best suited to their wants, and conceived that the superintendence of such institutions might be very properly associated with that of vaccination. The aversion or at least indifference of the 'natives' to the prophylactic, Dr. Smith considered, was of similar origin and character with their disfavour of European medicine. The Hindus and the Muslims still employed 'inoculators who produced smallpox, instead of vaccinators, who prevented it'.[31]

The idea of establishing public hospitals or dispensaries in the centre of the large and populous towns under the professional management of native surgeons and physicians educated in the rising Calcutta Medical College had for some time engaged the attention of the Earl of Auckland. By means of this agency, Auckland expected not only the immediate and effective relief of the 'native poor' from dangerous diseases by which the country was so often ravaged, but also a gradual, silent yet certain diffusion among the people at large of a 'correct knowledge and due appreciation of the benefits of medical science in its extended application to the preservation of public health, the protection from disease, the cure of sickness and prolongation of life'. Following out 'the expansive and enlightened' views of Lord William Bentinck regarding the medical education and subsequent employment of Bengali youth, the Governor-General began to entertain the conviction that the time had arrived when he 'might confide the management of a few such dispensaries substantially and responsibly into hands of some of the most distinguished young men who had taken their degrees at the Medical College at Calcutta'.[32]

Of all the Indian provinces, the Madras Government was probably the most responsive in its awareness of medical problems and active in taking measures to deal with them. One of the most important of these was the control of smallpox.[33] Unusually, vaccination against smallpox was compulsory in the Madras Presidency, and was carried out under a meticulous programme. Vaccine lymph was produced at the King Institute, local calves being used for the purpose. The calves were vaccinated on the belly and then confined in a special large byre under strict sanitary condition, looked after by special staff whose sole duty was to keep the animals and byre clean. If a calf defecated or urinated, the mess was immediately cleaned up. After a period of days the calf would become ready for collection of the lymph. This was collected from under the skin of the belly and preserved for use and the calf returned to the source from where it had been procured, none the worse for its experience. The returned calves were often seen in the bazaar and could be recognized by scar lines on the belly.[34]

A special staff of vaccinators then took up the job of vaccination. Here a very strict routine was carried out. A vaccinator was allotted a certain itinerary, going daily from village to village. He had to keep to a very strict schedule so as to be in a particular village on a particular day. On arrival he would vaccinate all babies not previously vaccinated and also those requiring revaccination, careful records being kept. The visits of operators were followed by inspectors who checked on the vaccinations carried out. This routine ensured that there was no covering up of inadequate work by incorrect reports. In this way, it is claimed that about two million vaccinations were carried out each year, and Colonel Short of Indian Medical Service believed that as a result, the people of the Province of Madras were 'probably the most universally vaccinated in the world'.[35]

Bengal presented a different picture altogether. Partly because of the lukewarm response of the government and partly due to popular inhibitions, vaccination was never adequately conducted in the province. Even in Britain itself, the benefits of Dr. Jenner's discovery were hardly appropriated. Sir J.Y. Simpson, the President of the Public Health Section of the Social Science Association, Edinburgh, estimated in 1868 that there still died about 5,000 annually from its ravages.[36] Some among these 5,000 had been duly vaccinated, and yet were susceptible to smallpox after cowpox just as men formerly were found susceptible to a second attack of smallpox after they had passed through a previous attack of natural or inoculated small-pox. Others seemed susceptible in consequence of the vaccination having been performed inadequately with imperfect matter. Again, a large number of those who perished from smallpox consisted of persons who had not been vaccinated at all, or who happened to have been exposed to the variolous poison antecedent to the age at which vaccination was usually performed. Doubtless, Colonel Simpson argued, 'a stricter enforcement of the new compulsory laws of vaccination, and a greater amount of attention to its proper performance with proper matter' would gradually diminish the number of the susceptible classes.[37] Coming to the situation in Bengal, the remedy for the evils, one can argue, perhaps lay in the temporary introduction of the Compulsory Vaccination Act in refractory areas, the appointment of paid government vaccinators and superior classes of men who could exercise greater influence on the people as Inspectors and Sub-inspectors of vaccination, and the more extended supply of good calf lymph.[38] But unfortunately such measures were never taken.

Having failed in their attempt to convince the Bengalis of the imperatives of vaccination against smallpox, the British health officers advised their younger colleagues to help themselves.[39] It was well, they argued, to be revaccinated against smallpox before proceeding to India. During residence in India vaccination should be repeated at intervals of five years. It must be remembered, they warned, that 'in India the infection of smallpox is constantly present'.[40]

One may recall the arguments of Sanjoy Bhattacharya, Mark Harrison and Michael Worboys in their work about the process of vaccination.[41] They argue that during the colonial period the government sought to control smallpox at four administrative levels: central, provincial, district and local. These multiple levels of administrative intervention failed to work in tandem; very often they worked against any uniform and concerted action. British officials, for instance, had to depend on their Indian counterparts at the local level to carry out the vaccination policy. But these Indian officials were often found to be empathetic with the community sentiments and therefore not so keen about enforcement. The policy also foundered on flaws of technical inadequacies. Apart from the problems of refrigeration of the serum, which in itself posed a serious hurdle, the technique of making a deep incision into the skin using a scalpel made the vaccination unpopular. However, the introduction of safer vaccine technologies gradually improved vaccination rates, and, during the 1930s, improved the public's perception of vaccination.

There were, however, some conscientious Indian health officials and educated lay persons in Bengal who, having some working knowledge of Western medicine, tried hard to impress upon the people and even the indigenous inoculators certain golden rules of vaccination. For instance, Gopal Chandra Majumdar, the Chief Assistant in the Vaccination Superintendent's Office in Calcutta argued in 1872 that people were extremely suspicious about the nature of modern vaccination, for they hardly knew anything about it. The vaccinators should, therefore, first dispel their doubts and then administer the vaccination. They should meet the village headmen or the agents of the zamindars and seek their help in convincing other villagers on the efficacy of vaccination.[42] One Ramanarayan Vidyaratna Bhattacharya wrote a treatise on vaccination in 1857 in which he argued that smallpox, a scourge in Bengal, carried away every year a large number of Bengalis. The ancient Hindu physicians had recommended for *tika* or modified smallpox through inoculation by the tikadars, but the remedy prescribed, Bhattacharya argued, was not sure, and not even lasting. People should therefore try out, Bhattacharya insisted, the Western method.[43] Haradhan Vidyaratna Kaviraja, still another practitioner of indigenous medicine, wrote in 1868 that treatment of smallpox by *vaidyas* had become outdated and consequently people had to fall back upon divine therapy.[44] But such voices were rare, and protest against vaccination went on for years together. Indigenous *tika* or inoculation remained the preferred treatment.

THE DISEASE DEITY AND THE *TIKADAR*

From very early times, we know, there had existed in India the worship of a goddess of smallpox, Sitala, and among certain castes it is still a rule to take no precaution whatever against the disease, a visitation of it in their households being regarded as a sign that they are favoured by the goddess. Smallpox was a direct intrusion of the goddess into the body of her victim. She expressed her presence in the patient's body in the form of fever.[45] If worshipped she would prevent the disease, but if provoked, the patient would perish.

But if familiarity with this dreaded disease evoked its ritualization, David Arnold believed, it also prompted attempts at medical control.[46] For more than half a century after the introduction of vaccination in 1802, nearly all the inhabitants in some parts of India were wedded to the practice of 'smallpox inoculation', and many thousands of Brahmin inoculators, with great influence over the people, were violent opponents of vaccination.[47] Partly because of the religious inhibitions and partly because of the large expenses involved, it had been extremely difficult for Bengalis to employ skilled medical agencies in the actual work of vaccinating and for many years the only vaccinators who could be obtained were illiterate and very ignorant. The purdah system as also the lack of female vaccinators posed a serious problem.[48]

Certain kinds of rumours spread among the urban poor. During the summer of 1878, for instance, when smallpox was going about in Lahore, John Campbell Oman was told by one of his servants that 'the Sarkar had given orders that the native women should be vaccinated, not as usual on the arm, but on the breast'.[49] From time to time indeed prejudices arose and rumours were spread in rural areas. A very early prejudice against vaccination arose because the vaccine came from the cow, an animal highly revered by the Hindus. One of the rumours causing difficulty in early years was to the effect that the object of vaccination was to set the 'government mark' upon the people who would afterwards be sent as coolies to other British possessions. Another rumour was floated that the government mark was a means of obtaining a census of the people with a view to imposing a new capitation tax.[50] Much more incredulous rumours were afloat for many years. It was a ploy to introduce plague; it was a well-calculated attempt at proselytization; it was intended to sacrifice vaccinated people for ensuring construction of a bridge or a railway embankment.[51]

We are told that in 1850 there were at least one *tikadar* for every 8 or 10 groups of houses, and that in Calcutta town alone the names and addresses of 68 inoculators were known in that year. From 1848 to 1867 attempts were made by various civil surgeons to estimate the extent to which smallpox inoculation was practised in Bengal. From an examination of more than 35,000 people in those years it appears that over 81 per cent had been inoculated. At least two accounts of the mode in which the operation was performed by professional inoculators are available: one by R. Coult in 1731, and the other by J.Z. Holwell in 1767. Francis Buchanan has also elaborated information on the practice and its practitioners.[52] All of them did concur that inoculation consisted of matters originally taken from the pus of smallpox pustules with a thick needle, but they disagree on the process of administration. While Coult argued that the pus was taken from a patient when the pox came to maturity, Holwell believed that it was taken from persons who contracted it in a milder form from the inoculation itself and was then preserved in cotton within a rag for a year. Buchanan says it was only taken from pustules of those who had contracted the ailment 'naturally', and kept in cotton for no more than three days. Again, while Coult believed that punctures were made with a needle impregnated with the pus, the *tika* being laid on the upper arm, Holwell argued that the incisions were made on the outside of the arm by a barber's instrument for cutting nails, and the variolous matter was secured by placing it in a moistened cotton-wad put on the cut for hours. Buchanan corroborated Holwell that first the incisions were made and then the variolous matter soaked in the cotton was rubbed on them.[53]

There is again little agreement in the contemporary evidence about who the inoculators were. Holwell says, 'inoculation is performed in Hindoostan by a particular tribe of Brahmins,' who came on regular tours from Brindaban, Allahabad, Banaras, and 'arrive commonly in the Bengal provinces early in February.'[54] Buchanan believes that the inoculators were 'of both religions (i.e. Hindus and Muslims) and of all castes'. *Tikadars* or mark-makers also hailed from marginalized caste groups such as *napits* or barbers, *malis* or gardeners, cultivators, garland-makers, *tanty* or weavers, *kumhars* or potters.

Both Holwell and Buchanan believed the success rate to be high. To Holwell, it was 'a miracle to hear that one in a million fails of receiving the infection', from the inoculation. And Buchanan testifies that when smallpox struck district Dinajpur as an epidemic it could take as its tolls only one of a hundred of those inoculated. J.Z. Holwell for his part gives a detailed story of how the feat was performed:

The inhabitants of Bengal, knowing the usual time when the inoculating Brahmins annually return, observe strictly the regimen enjoined, whether they determine to be inoculated or not; this preparation consists only in abstaining for a month from fish, milk, ghee. . . . When the Brahmins begin to inoculate, they pass from house to house and operate at the door, refusing to inoculate any who have not, on a strict scrutiny, duly observed the preparatory course enjoined them. They inoculate indifferently on any part, but left to their choice, they prefer the outside of the arm midway between the wrist and elbow, and shoulders for the females. Previous to the operation the operator takes a piece of cloth in his hand and with it gives a dry friction upon the part intended for inoculation, for the space of eight or ten minutes; then, with a small instrument, he wounds by many slight touches, about the size of a silver groat, just making the smallest appearance of blood. Then opening a linen double rag (which he always keeps in a cloth round his waist) he takes from thence a small pledget of cotton charged with the variolous matter, which he moistens with two or three drops of the Ganges water, and applies it to the wound, fixing it on with a slight bandage, and ordering it to remain on for six hours without being moved; then the bandage to be taken off, and the pledget to remain until it falls off itself. The cotton, which he preserves in a double calico rag, is saturated with matter from the inoculated pustules of the preceding year; for they never inoculate neither with fresh matter, nor with matter from the disease caught in the natural way, however distinct and mild the species. Early in the morning succeeding the operation, four *collons* of cold water are ordered to be thrown over the patient from the head downwards, and to be repeated every morning and evening until the fever comes on, which usually is about the close of the sixth day from the inoculation; then to desist until the appearance of the eruption (about three days) and then to pursue the cold bathing, as before, through the course of the disease, and until the scabs of the pustules drop off. They are ordered to open all pustules with a sharp pointed thorn as soon as they begin to change their colour, and whilst the matter continues in a fluid state. Confinement to the house is absolutely forbid, and the inoculated are ordered to be exposed to every air that blows; and the utmost indulgence they are allowed, when the fever comes on, is to be laid upon a mat at the door. But in fact the eruption fever is generally so inconsiderable and trifling as very seldom to require this indulgence. . . . Their regime is ordered to consist of all the refrigerating things the climate and season produce. . . . These instructions being given, and an injunction laid on the patient to make a thanksgiving *poojah* or offering to the goddess on their recovery, the operator takes his fees, which from the poor is a pun of cowries, equal to about one penny sterling, and goes on to another door, down one side of the street and up on the other; and is thus employed from morning till night, inoculating sometimes eight or ten in a house.[55]

I have quoted this long passage to underpin certain crucial points. It can also be seen from the contemporary accounts that the operation was carried out with great care, and there can be no doubt that in those comparatively olden times a high degree of knowledge about the procedure necessary for success had been attained. In this connection, certain other important details regarding the practice as carried out in that period are worth mentioning.

First, infants at the breast were not inoculated, but children above one year of age were considered old enough for the operation.

Second, the rules to which in accordance with ancient custom everyone desiring inoculation had to conform were not only arduous in themselves but interfered greatly with daily work and business. The restrictions included abstinence for a month prior to inoculation from milk, fish, and ghee. In addition to that, for 21 days after the operation, no member of an inoculated household was permitted to have contact with the outside world, and no person from another village was permitted to enter a house containing inoculated patients. Each time if either of these rules was infringed the guilty person was regarded as 'unclean', and was required to bathe and to put on different clothes.

Third, the early professional inoculators freely acknowledged the infectiousness of the disease set up by inoculation, and the elaborate code of rules to which the people in inoculated villages had to conform, bears abundant testimony to this. Thus, no inoculation was allowed in a village unless nearly all the unprotected people were willing to have the operation performed; all the inoculations in a village were done on the same day; women who were pregnant and others who could not arrange to be inoculated on the date had to leave the village until the danger of infection was past; the clothes of all inoculated persons were kept for 21 days and then washed on a fixed day separately from those of unprotected people; no barber was allowed to ply his trade in any house where inoculated persons resided; bathing in public tanks was not permitted until the eruption had disappeared.

Finally, the fatality attending the operation, depending as it did very greatly on the skill and care of the operator, was probably very low in the period to which we are referring. This is evidenced by Holwell himself, and much later by the observations of Bedford who saw 79 inoculations performed without a death resulting, and Wise who reported that in Chittagong the mortality attending the operation was only 0.5 per cent.[56]

We see then that in olden days when all the rules enumerated were strictly adhered to, and when only the professional Brahmin inoculators performed the task, the measure proved to be a real blessing to the inhabitants of certain parts of India, especially Bengal. We must also remember that the conditions of certain parts of India where the operation was general were very favourable to the success of the measure. The great bulk of the inhabitants of Bengal, Assam and Burma lived in small villages, hamlets, each house being some distance from the next, and each village, hamlet a mile, or more from neighbouring ones. Under such conditions, and with the strict rules of segregation in force, the danger of spreading the disease set up by inoculators was not great.

Unfortunately, there remains another side of the story. With the increasing belief in the efficacy of the measure, the demand for inoculators became greater than the supply and the price charged for the operation rose. In 1850 an usual fee was 4 *annas* and some rice for the operation on a boy and 2 *annas* and some rice for a girl;[57] other writers of the same period say that the lowest fee taken for puncturing was Rs. 2 or 3 and the highest Rs. 16 and a shawl. As time went on, the work of inoculation was undertaken by Hindus of low castes or people who were not only exceedingly ignorant but had not the influence necessary to make the people conform to the rules for preventing the spread of the disease. The result was that about the middle of the nineteenth century many epidemics were traced to inoculation. In 1831, Dr. Cameron stated: 'It is now well ascertained that inoculation is the great means by which smallpox is kept in existence in Calcutta', and in 1844 the Superintendent of Vaccination reported, 'Smallpox is annually introduced into Calcutta by a set of inoculators, numbering about 30, to the great endangerment of the public health'.[58]

By 1850, it was ascertained that in Bengal their number had risen to 68. It appears that 42 of them were permanent inhabitants of Calcutta, residing chiefly in the northern division or 'native part' of the town. They were principally of low Hindu castes or trades, such as those of *malee*, *tanty*, *kumhar*, and *napit*, from which they derived their principal livelihood during nine months of the year, practising their profession only during the other three. The rest were Brahmins or *doibogyas*. They were said to have come to the town annually from Burdwan, Hooghly, Birbhum, and the adjacent west and northern districts, generally in the month of *Magh* or February, remaining in town for about 4 or 5 months, and inoculating on an average 70 or 80 persons each day. It appears that very poor people used to pay about Rs. 2 to the *tikadars* for 2, 3, 4, or even 5 children in a group or family, which was the custom. Middle class people paid from Rs. 3, 4 to 10 for a similar party, and much more to the Brahmins for the performance of *puja*, the most expensive part of which was the *gaun* or canticle to be sung for several successive days in honour of Sitala.[59] When the operator was a Brahmin himself, or rich enough to possess an idol of the goddess, the *puja* was performed at the nearest shrine of the deity, the fees were in that case paid to the officiating priest there.[60]

Evidence available indicates that expenses actually incurred by a respectable middle class Bengali for inoculating three children amounted to Rs. 15 and 4 *annas*.[61] This included expenses for the actual act of inoculation, *puja* for 15 days performed at home, cotton clothes, gold, sweetmeats, and priest's fees.

At times, the inoculators increased their income largely by treating cases of smallpox so that it was to their pecuniary advantage when epidemic broke out. Indeed, it was said that when they reached a village, where, in consequence of the absence of smallpox, the people were reluctant to be inoculated, it was not unusual for them to throw pieces of cotton soaked in smallpox matter in places where children played, knowing fully well that if they could conjure up an epidemic, people would flock to them for inoculation. No one would, therefore, be surprised to see that in the practice of inoculators of this period, the mortality attending the operation was often very high. Thus, the Civil Surgeon of Serampore reported in 1850 that one inoculator admitted having treated during the season 400 individuals of whom 200 died.[62] It was stated publicly in the *Bhaskar*, a popular Bengali newspaper, that around 1850 in the villages of Sonatikoree in Hooghly district nearly 1,000 boys and girls were inoculated, of whom 300 died. Again in the village of Caderpore among 100 children who were inoculated, more than 20 died.[63]

Many in Bengal sought to prohibit the practice of inoculation since it supposedly led to outbreak of smallpox epidemics if the rules of inoculation were not strictly followed. Some of the *bhadralok* in Calcutta including Pundit Modusoodan Gupta, Coomar Kally Kissen Ray, Issurchunder Singh alongwith Mr. G. Naylor, Sub-Assistant Surgeon of the Gurranhattah Dispensary, and many other Civil Surgeons believed that inoculation was more often than not fatal.[64]

In France the practice of inoculation was prohibited by law as early as 1793, but in England it required the epoch-making discovery of Jenner and over forty years experience of the benefits of this discovery before prohibition of inoculation was passed. In india, as early as 1804, an attempt was made at the desire of Wellesley to prohibit smallpox inoculation within the town of Calcutta. The observance of the rule was not, however, made compulsory by law and it gradually sank into neglect. It was not until 1865 that smallpox inoculation was prohibited by law in Calcutta and its suburbs, the Act being extended in the following year to villages in the neighbourhood of Calcutta and to several large stations in Bengal proper. Until then people used to die of smallpox in large number partly because of shoddy handling of inoculation by inexperienced and greedy *tikadars*.

At shrines of Sitala prayers were generally offered and vows made by the relatives of the afflicted. She was worshipped by almost all classes, but she seems to have got the most patronage by the poorer section of the society. She is prone to anger and quick to punish anyone who disturbs the natural balance of hot and cold in his body.[65] It was popularly believed that when this occurred she was provoked, and she would punish the person and leave her mark on his body. The pock mark she thus left on the infected body would be referred to by people as *mayer daya* or the grace of the mother.[66]

Duncan Stewart records that he happened to possess a 'curious' ancient gold coin about half an inch square, which he believed represented the goddess Sitala in her usual attire.[67] She is seated naked upon an ass bedecked with flowers and jewels. She holds in one hand a broom, representative of the duty of cleanliness. Resting on the left hip and supported by that arm is a large water jar indicative of the necessity of ablution and coolness. On her head she carries a winnowing sheaf or fan to be used as a ventilator. Medals of the same description of gold or silver were in fact worn by people as amulets. But it appears that the essential duties of observance of cooling, cleanliness and proper ablution enjoined on the worshippers of Sitala were neglected for the more important priest craft, incantation ceremonies, and propitiatory offerings.

Contemporary Bengali dignitaries too expressed an almost similar opinion. Baboo Russomoy Dutta, for instance, argued that there was hardly any medical treatment at all; the cure relied instead on the mercy of Sitala. A passage in the prayer usually offered to Sitala runs: 'No charms or medicines exist for the wicked disease'. The usual mode of treatment, as evidenced by Dutta, was as follows.[68] In the first stage of five days, when the patient was labouring under fever, he was almost starved; only dry sweets or sugar candy (*batassa*) was given for food, and *jaree* and *panchan*, were administered as medicines. In the second stage after five days, when the fever went down, food was given: milk and rice, rice and *moog jurie*, and milk *mooky* after the eleventh day. On the seventh day, cold water was sprinkled over the patient's body, and on the ninth day, a thin poultice made of turmeric, linseed meal, and milk was applied over the pustules. On the twelfth day, the diet was restricted to thin chapatti, *moog jurie* and fried potatoes, *potol* and *katchkala* (green plantains) pottage/broth, and that too in small quantity. This continued up to the fifteenth day, when generally the patient was expected to recover. Almost all the practicing physicians in Bengal used to prescribe this sort of food and diet, but all of them ruled out any sort of medical intervention.[69]

Dutta also enumerated certain features of smallpox in Bengal. First, all classes and castes of Bengal, all ages and both sexes, he argued, were equally susceptible of infection. Mortality was mainly dependent upon the modifying circumstances of previous inoculation or vaccination. Nevertheless, it also depended on certain other conditions of the victim: (a) the natural constitution, (b) present health, (c) personal comfort, and (d) poverty affecting the individual. Mortality, Dutta argued, depended also on the particular constitution of the atmosphere, the locality, and the construction of the dwellings.

Second, the great majority of the victims were totally unprotected either by vaccination or previous inoculation, though the latter practice was most common.

Third, those who had had the disease previously, or who had been *successfully* vaccinated, always had the disease in a *modified* form. The incursive fever was often equally violent, but the eruptive stage was always milder, and the secondary fever proved fatal only in previously debilitated or scrofulous subjects.

Fourth, in those who suffered from high fever at the onset with much cerebral and nervous excitement, headache, delirium, and severe lumber pain, the eruptive stage succeeded most fully and favourably, affording though in a confluent form, great relief to the sufferings and promise of a favourable termination. In these cases the chief danger used to arise on the twelfth and fourteenth day from secondary fever.

Fifth, the worst and most certainly fatal cases were those of poor and weak patients. Finally, a damp and hot atmosphere, its diffusion promoted by dry and cold weather moderated the violence of the epidemic.[70]

Dr. H.H. Goodeve of Calcutta Medical College argued in the 1840s that the purely European population of the city was peculiarly exempt from the true smallpox during the epidemic of the 1830s. A few cases did occur but those were chiefly in the lower classes. Goodeve neither saw for himself nor had he from others heard of more than a dozen cases in the respectable portion of the community. This exemption, Goodeve believed, was caused by better houses, cleanliness, ventilation, and food, but there was no doubt it might be attributable also to the greater care taken to give full effect to the operation of vaccination in childhood.[71]

CONSTRUCTING THE EPIDEMIC

These were evidently the discourses of the literati. But the moot question is what was the *popular* approach to health in Bengal as evidenced by the outbreak of smallpox in the early nineteenth century? The focus was arguably not on microbes, vectors, antigens and antibiotics, but on right conduct, over-indulgence, sin, and the intervention of demons and deities. Popular health culture in Bengal, it has generally been argued, enshrined an amalgam of Ayurvedic, religious, folk 'magic' and other elements. Western medical intervention was limited to a small, educated minority.[72] We have already noticed that British officers of the nineteenth century generally tended to believe that the diffusion of smallpox was largely due to inoculation by the *tikadars*. It is difficult to trace the origin of this practice in Bengal, but there is hardly any doubt that it is of great antiquity, and that it was extensively practised by all classes of people in Bengal, Hindu and Mohammedan. From investigation done by Dr. Wilson of Bauliah, Dr. Wise of Dacca and others, it appears that in all probability 70 per cent of the population of Bengal was inoculated by the middle of the nineteenth century.[73] There had existed a popular belief, industriously supported by succeeding generations, that the observance of this practice by Hindus at some period of life was a religious duty, and its neglect was criminal or at least disreputable. However, traditional pundits in Bengal seemed to be a confused lot, not categorically in its favour, nor decidedly against it.

The Committee on Smallpox Inoculation, 1850, instituted a most searching enquiry on this matter. It submitted several questions on smallpox inoculation to different pundits. First, is there in the holy *Shastras* any distinct commandment enjoining smallpox inoculation as a religious duty, or recommending it as a commendable act? Second, what religious or other observances are enjoined on those who are attacked by smallpox? Third, was the avoidance of inoculation considered a sin, or disreputable? Fourth, is there in this world any penalty for the omission or punishment in the next, and if so, how may it be atoned for? Fifth, supposing it is proved that the vaccine disease is really a modified form or variety of the smallpox, should not all the religious ceremonies observed hitherto in cases of common smallpox be attended to after vaccination by good Hindus?[74]

Replies to these questions by the pundits of Nadia, those of the Banaras College and of the Sanskrit College of Calcutta, and the pundits of the Court of Sudder Dewanny Adalat were quite unanimous in declaring that although certain religious observances were proper or incumbent on all persons being Hindu who contracted smallpox in any form, there was no sort of obligation, moral or religious, on any one to subject himself or his children to inoculation. They further argued that neither penalty nor reproach was attached to its omission.[75] Banaras pundits argued:

There is no reference to Inoculation in the Holy Books, but if the practice be calculated to save life, it must deserve to be praised.... On the appearance of smallpox, the Goddess Sitala is to be worshipped by recitation of praises and feeding of Brahmans... As neither praise nor blame is assigned to the practice of inoculation in the Shastras, so neither in the practice in common life regarded either with approval or disapproval.... In consequence of the absence in the Shastras of any injunction as regards inoculation, of course there is no penalty with reference to it.... If it is ascertained that there is no essential difference between the ordinary smallpox and the pustular disease which follows vaccination, of course the ceremonies proper in the one case are proper in the other.[76]

A more guarded reply to the same queries came from the Sanskrit College Professors and Pundits, and runs thus: 'In case of *modified smallpox* in a person who has been punctured in the English manner, the worship of Sitala may be performed with all the usual formalities, she being the guardian Deity of Smallpox Disease; such religious observances are necessary.'[77] But the most definitive answer came from the Pundit of the Sudder Dewanny Adalat in Calcutta: 'The Vaccine Disease is in reality a modification or variety of smallpox; such being the case, virtuous Hindoos when vaccinated, should with veneration, faith and purity, observe the same religious ceremonies mentioned in my answer to the fourth question as they do when attacked with smallpox.'[78] It was necessary for victims of smallpox to live in secluded, pure, cool, and pleasant rooms, free from impure things such as remnants of food, and to keep themselves clean. It was also necessary for them to hang leaves of *neem* trees (*Melia azaddiracta*) around the rooms; to worship Sitala Devi according to their ability, with devotion, faith, and purity; to chant her name frequently; to make *homa*, to present offerings, to give alms, to perform rites for averting the evil, to respect Brahmins, worship cows, worship Shambhu and Gauri (Lord Shiva and Goddess Durga), and cause faithful Brahmins to sing hymns to Sitala—all these were necessary to be done by the patient for the alleviation of the pain of smallpox.[79]

Nevertheless, indigenous practitioners and educated persons were more often than not in favour of vaccination. Even in the early nineteenth century attempts were made to overcome the popular prejudices to vaccination by resorting to a 'pious fraud' that pre-Jennerian smallpox vaccination was practised in India.[80] We are informed that Mr. Ellis of Madras who was well versed in Sanskrit composed a short poem on vaccination in that language. 'This poem,' the story goes, 'was inscribed on old paper, and said to have been found, the object of the pious fraud being that the impression of its antiquity might help to reconcile the minds of the Brahmins to the use of a prophylactic drawn from their sacred cow.'[81]

Again, Calvi Virumbon, a Hindu, is said to have adduced reference to a supposedly ancient passage in support of Western vaccination. Since vaccine prepared from cowpox was revolting to Hindu perception, Virumbon composed a Sanskrit verse to legitimize the introduction of Jennerian vaccination. The Sanskrit verse enjoined: 'take the fluid of the pock on the udder of a cow…then mixing the fluid with the blood, the fever of the smallpox will be produced'.[82]

Even earlier, on 29 December 1804, one 'obedient and very humble servant' Mooperal Streenivasachary wrote a letter to Dr. J. Anderson stating that 'it is therefore greatly to be wished that an intimate knowledge of this wonderful discovery may be acquired by the natives of this country, so as to enable them to preserve the lives of the rich and honourary as those of the low caste'.[83]

Such persuasive passages were in fact not wanting in Bengal. In 1857, one Ramnarayan Vidyaratna Bhattacharyya in fact wrote a treatise on vaccination arguing that the Western method of vaccination was more scientific and effective.[84] This was corroborated by Haradhan Vidyaratna Kaviraja who in 1868 observed that inoculation had almost ceased to be practised in Bengal and that the Bengalis should go in for vaccination for their own health interest.[85] Generally speaking, the prevalent attitude of the people in Bengal towards vaccination was one of passive acceptance of the prophylactic.[86] But in nearly every village, there were certain families who habitually refused vaccination either openly or by covertly hiding away their children from the vaccinators. This was due not so much to the vaccination itself as to the use of human vaccinifers. What the villagers seemed most to object was having to pay for the operation and the using of some of their children as vaccinifers. For, this was attended by the discomfort of having to carry their children to the neighbouring village.[87] Moreover their principal objection was that it was done without *puja* or any kind of sacrificial offering.[88]

There were, however, many doubts even among the educated Bengalis about the effectiveness of the method of vaccination. 'Many people', writes a Bengali medical journal in 1894,

are taking vaccinations for fear of small pox, in some cases this is yielding results, in some cases the vaccines are being rejected by the body, and in others, those who are being vaccinated are falling prey to pox as soon as they are vaccinated. However, *those who have not yet got themselves vaccinated should better not take it now, for quite a few people have succumbed to small pox after taking the vaccine meant to prevent it* (emphasis added).[89]

They also sounded some words of caution as preventive measures:

It has been heard that *koi* and *magur* fish have also been infected by pox. Those who fear the disease must not eat these types of fish. If you wear the seed of *horitaki* as a charm around your arm, then small pox cannot attack you. The medicine mentioned below has been advocated in the *Hitabadi*: 2 raw roots of the *kantikari* plant, ground with 2½ peppers, if eaten, will ward off small pox for a lifetime. For those who already have the disease, eating this preparation will free them from it. Wounds will heal soon if the fat of the frog is applied on them. Thousands of people have been cured by this type of treatment.[90]

The objection to vaccination can perhaps be explained in four ways: first, the very great prejudices of the Bengalis from the highest to the lowest to adopt anything contrary to or deviating from the established customs of their ancestors; second, the inoculators were principally Brahmins of high caste, who enveloped in 'mystery' the process of inoculation, bringing in religion to their profession, i.e. Sitala must be propitiated, *puja* should be performed and offerings made; third, to get vaccinated with lymph drawn from a cow was to the Hindus nothing less than eating beef;[91] four, the Hindus were impressed with a belief that *Basanta* or smallpox was above all medical care, for the indigenous medical works prescribed no remedy for it, and that a patient was entirely at the mercy of Sitala, whom they considered it their duty to propitiate by worship and various offerings, the *tikadars* acting in the capacity both of doctors and priests on such occasions.[92] The Hindus who most resisted vaccination were the Brahmins, Marwaris, Rajputs and Banias. As years rolled on, some progress undoubtedly occurred in the popularizing of vaccination, and people came to regard it as the government *dustur* or order to have the operation performed. To a certain extent, they then seemed to appreciate its benefit.[93]

Among the Muslims, the Ferazis perhaps gave the most opposition.[94] They successfully withstood every attempt, whether by the Sanitary Department or local officials, to introduce vaccination among them.[95] The Ferazis belonged to the Sunni division of the Mohammedan faith, and might aptly be described as the Puritans of the Islamic religion, their position being very similar to that of the Pharisees in the Jewish dispensation. One Haji Shariatulla, who went to Mecca and Medina and came under the influence of Abdul Wahab, established the sect. On return to India, he settled in Daulatpur village in Faridpur district. Finding that the religious life of the Muslims had greatly deteriorated by the contact with Hindu idolatry, he addressed himself to the task of reforming their practices with the aid of the new theology he had learnt in Arabia. In fact, it was a modified Wahhabism that he sought to enforce, and it was not long before he gathered an immense following in Eastern Bengal.

Shariatulla taught *inter alia* that inoculation and/or vaccination should be forbidden on the following grounds: 1. that it being the custom among Hindus and degenerate Muslims to do *puja* to the goddess Sitala before being inoculated, and all connections with Hindu worship being forbidden, inoculation should be rigidly proscribed; 2. that no prophylactic measures should be taken against disease; 3. that it was a sin for the Muslims to have pus or blood from man or animal injected into their bodies.[96] Maulavi Sharitulla's following increased, and by the end of the nineteenth century there were over one million Ferazis in Shibchar thana alone.

Clinging as they did to their religious prejudices with great tenacity, the Ferazis had resisted vaccination in a most determined spirit. They drove inspectors, sub-inspectors of vaccination and vaccinators away from their houses until those officials became afraid to venture into their villages.[97] One of their great objections to vaccination was a belief that they had among them an Imam Mehndi, who was to be their future ruler. One of his peculiarities, they believed, was that he had no blood in his veins, and that he had milk in his veins instead. Aware of this and wishing to seize him, Government had instituted vaccination as a ploy by which he might be identified, because the man from whose punctured arm milk flowed, instead of blood, must be Imam Mehndi! As soon as he was discovered he would be taken away and handed over to the authorities. This 'tale' was firmly believed by the villagers, and stood in the way of their submitting to vaccination.

There was the same opposition to vaccination in Bihar and in the districts of Orissa even in the opening years of the twentieth century.[98] In fact, there was scarcely a district in Bengal which failed to register some form of opposition. Vaccination was indeed very unpopular in most parts of the province. Parents used to hide their children from the vaccinators in the belief that the operation would cause them pain and lead to suffering and disease. Not infrequently, deaths from other causes, months after children were vaccinated were attributed to vaccination. Resistance to vaccination was even stronger in Madras. In 1804 Surgeon J. Dalton was accosted by an angry crowd who refused to submit to vaccination and declared that they would prefer death to vaccination.[99] More hostile to vaccination were the Mapilla Muslims in Malabar, not unlike the Ferazis in Bengal.

The vaccinators, therefore, had to work under great pressure. They were not government servants and were not recognized as such by the people. In the absence of a law to enforce vaccination, the Vaccination Department had to depend chiefly on persuasion to advance the cause. But persuasion was of no avail with poor people who had not the wherewithal to pay for the bare necessities of life and could ill afford to pay for vaccination, however, beneficial it might be. The result was that the vaccinators left 'unprotected' the areas which they considered least profitable and carried their work in places where they got their fees and sometimes even an article of food. The inspecting staff that for the most part were promoted vaccinators were not men of sufficient social status to be able to inspire confidence. They often shared the spoils of the vaccinators and attempted to protect them.[100]

There were thus numerous families that neither inoculated nor vaccinated their children. Whenever one of them had been naturally attacked by smallpox, they considered that the goddess of smallpox Sitala had been propitiated and would leave the others alone.[101] Thus they remained from generation to generation uninoculated and unvaccinated in that family, however, it might be increased. Evidence available indicates that this was the case generally with the middle and lower classes of the Hindu population in towns and villages. The Hindustani people were, however, most bigoted in regard to Sitala, and they believed that if *Matta* or smallpox came naturally into a family, it was favourable and good. In this belief they did not allow themselves to be inoculated nor vaccinated, and therefore the disease generally attacked them in a very severe form.[102]

CONCLUSION

A few tentative conclusions can be drawn from all these. Popular responses to epidemic smallpox in nineteenth century Bengal offer a mixed and complex pattern, and these complexities stemmed from varied and differential perceptions of the epidemic. Generally speaking, the response of the masses was one of stoic acceptance of the disease and not a dreaded repulsion.

Coming to the question of vaccination, the event of its acceptance was not a gradual process; nor was resistance uniform. The metropolitan or educated Hindus/Brahmins were gradually accepting the value of vaccination as a remedial measure. Early years of the nineteenth century witnessed the formative period of this transformation from resistance to acceptance. By contrast, the unlettered, superstitious and tradition-bound village masses were slow to acknowledge its efficacy. Partly because of the 'pious fraud', but largely due to government persuasion/coercion during the second half of the nineteenth century, the low caste/Hindu masses relented and went in for Western method of vaccination.

Historically speaking, eighteenth-century India was punctuated by dramatic events which brought in profound socio-economic consequences for its people. Progressive revenue demands by the British government had pushed the peasants to extreme poverty and squalor; the English traders in Bengal vowed to buy cheap, sell dear and corner commodities. Poverty and malnutrition lowered people's resistance to infectious diseases. Recurrent Maratha raids and the Nawabs' battles with the Company had disrupted village life to the extent that the villagers' indigenous mode of disease control through the regular annual visits of the *tikadars* was thrown out of practice.

Regular diet and inoculation failed. All the villagers had been left with was their capacity to write and sing hymns to the goddess, which the socio-political turmoil occasioned by the colonial intervention could not destroy. Not only did the *tikadars* recite hymns to the goddess Sitala while inoculating and enjoined the inoculees to recite along with him, but the inoculees had to prepare themselves by abstaining from fish, ghee and milk. This procedure indeed constituted their subjectivity in a pre-modern social milieu.

NOTES

1. Anil Kumar, *Medicine and the Raj: British Medical Policy in India, 1835-1911*, New Delhi: Sage, 1998.
2. Poonam Bala, *Imperialism and Medicine in Bengal: A Socio-Historical Perspective,* New Delhi and London: Sage, 1991.
3. David Arnold, 'Smallpox and Colonial Medicine in Nineteenth-Century India', in David Arnold, ed., *Imperial Medicine and Indigenous Societies*, Manchester: Manchester University Press, 1988.
4. Kabita Ray, 'Smallpox and the Introduction of Vaccination in Colonial Bengal', in Abhijit Dutta et al., *Explorations in History*, Calcutta: Corpus Research Institute, 2003. Sanjoy Bhattacharya, 'Re-devising Jennerian Vaccine: European Technologies, Indian Innovation and the Control of Smallpox in South Asia', in Biswamoy Pati and Mark Harrison, eds., *Health, Medicine and Empire: Perspectives on Colonial India*, Hyderabad: Orient Longman, 2001.
5. Ralph W. Nicholas, 'The Goddess Sitala and Epidemic Smallpox in Bengal', *Journal of Asian Studies*, vol. XLI, 1981.
6. David Arnold, *Colonizing the Body: State Medicine and Epidemic Disease in Nineteenth Century India*, Berkeley, Los Angeles, London: University of California Press, 1993
7. Sanjoy Bhattacharya, Mark Harrison and Michael Worboys, *Fractured States: Smallpox, Public Health and Vaccination Policy in British India, 1800-1947*, New Delhi: Orient Longman, 2005
8. Harish Naraindas, 'Care, Welfare, and Treason: The Advent of Vaccination in the 19th Century', in *Contributions to Indian Sociology*, 32, 1, 1998, pp. 67-96.
9. Ralph W. Nicholas, 'The Goddess Sitala and Epidemic Smallpox in Bengal', *Journal of Asian Studies*, vol. XLI, 1981.
10. Deepak Kumar, 'Social History of Medicine: Some Issues and Concerns', in Deepak Kumar, ed., *Disease and Medicine in India*, New Delhi: Tulika, 2001.

11. B.J. Vyas, *Notes on Sanitary Primer*, Ahmedabad: no pub., 1886, pp. 19-20, Oriental and India Office Collection (hereafter OIOC), British Library (hereafter BL), London.

12. E.A. Gait, *A History of Assam*, Calcutta: Thacker, Spink & Co., 1906, p. 100.

13. *Voyages to the East Indies*, pp. 153-4, cited in *Extracts from Records of Past Epidemics in India*, 1912, National Archives of India (hereafter NAI), New Delhi.

14. J.Z. Holwell, *An Account of the Manner of Inoculating for the Smallpox in the East Indies*, London, 1767; reprinted in Dharampal, *Indian Science and Technology in the Eighteenth Century: Some Contemporary European Accounts*, Delhi: Impex India, 1971, p. 144.

15. Haradhan Vidyaratna Kaviraja, *Basanta Roger Nidan o Chikitsa*, Calcutta: no pub., 1868, p. vi. V T/ 1343, OIOC, BL, London.

16. Ibid., p. 1.

17. Ray, op. cit., p. 106.

18. Ibid.

19. *Annual Report upon Vaccination in the North-Western Provinces*, 1866-7, p. 4. S.P. James, *Smallpox and Vaccination in British India*, Calcutta, 1909.

20. *Report of the Smallpox Commissioners Appointed by Government with An Appendix, Calcutta, 1st July 1850*, Calcutta, 1850, p. 2, NAI, New Delhi, (hereafter *Report of the Smallpox Commissioners, 1850*).

21. Ibid.

22. Ibid.

23. Charles T. Edmondson, *Popular Information on Smallpox, Inoculation and Vaccination*, Calcutta: no pub., 1870, p. 1.

24. Ibid.

25. Drewitt Dawtrey, *The Life of Edward Jenner*, London: Longmans Green & Co., 1931, p. 63.

26. Sanjoy Bhattacharya, Mark Harrison and Michael Worboys, *Fractured States: Smallpox, Public Health and Vaccination Policy in British India, 1800-1947*, Hyderabad: Orient Longman, 2005.

27. Major H.J. Dyson, *Fifth Triennial Report of Vaccination in Bengal during the Years 1899-1900, 1900-01 and 1901-02*, Calcutta: Bengal Secretariat Press, 1902, p. 9, NAI, New Delhi.

28. Ibid.

29. Duncan Stewart, *Report on Smallpox in Calcutta, 1833-34, 1837-38, 1843-44, and Vaccination in Bengal from 1827 to 1844*, Calcutta: Government Press, 1844, pp. 191-3. V/27/857/2 OIOC, BL, London.

30. Ibid., pp. 209-10.

31. Ibid., p. 110

32. Ibid., p. 211.

33. Colonel H.E. Short, IMS, *In the Days of the Raj, and After: Doctor, Soldier, Scientist, Shikari*. MSS Eur C 435, OIOC, BL, London, pp. 133-4.

34. Ibid., p. 134.

35. Ibid.

36. Sir J.Y. Simpson, *Proposal to Stamp out Smallpox and Other Contagious Diseases*, Edinburgh, 1868. OIOC, BL, London, p. 4.

37. Ibid., p. 5.

38. Major H.J. Dyson, *Fifth Triennial Report of Vaccination in Bengal During the Years 1899-1900, 1900-01 and 1901-02*, Calcutta, 1902, p. 10, NAI, New Delhi.

39. Major General J.B. Smith, Medical Adviser to the Secy of State of India, *Simple Rules of Health for Young Officers Proceeding to India for the First Time*, India Office, 1926. L/PO/6/32, OIOC, BL, London, p. 5.

40. Ibid.

41. *Fractured States: Smallpox, Public Health and Vaccination Policy in British India, 1800-1947*, Hyderabad: Orient Longman, 2005.

42. Gopal Chandra Majumdar, *Tikadarganer Prati Upadesh*, Calcutta, 1872, pp. 7-8. VT/949, OIOC, BL, London.

43. Ramanarayan Vidyaratna Bhattacharya, *Gobeej Proyog, Arthath Gobeej Dwara Tika Dibar Bidhi*, Calcutta: no pub., 1857, p. 1. VT/1637(c), OIOC, B L, London.

44. Haradhan Vidyaratna Kaviraja, *Basanta Roger Nidan O Chikitsa*, Calcutta: no pub., 1868, p. iii. VT/1343, OIOC, BL, London.

45. Lawrence A. Babb, *The Divine Hierarchy: Popular Hinduism in Central India*, London: Columbia University Press, 1975, p. 277.

46. David Arnold, 'Smallpox and Colonial Medicine in Nineteenth-Century India', in David Arnold (ed.), *Imperial Medicine and Indigenous Societies*, Manchester and New York: Manchester University Press, 1988, p. 49

47. S.P. James, *Smallpox and Vaccination in British India*, Calcutta: Thacker, Spink & Co., 1909, p. vi.

48. *The Indian Medical Records*, vol. LII, Feb. 1932, p. 42.

49. Oman John Campbell, *Cults, Customs and Superstitions of India*, Delhi: Vishal Publishers, 1972.

50. James, op. cit., p. vii.

51. Ibid., p. vii.

52. *An Account of the District of Purnea in 1809-10*, Patna, Bihar and Orissa Research Society, 1928, p. 187; *An Account of the District of Bhagalpur in 1810-11*, Patna, Bihar and Orissa Research Society 1939, p. 173. See also Montgomery Martin, *The History, Antiquities, Topography and Statistics of Eastern India*, London: W.H. Allen and Co., 1838, II, pp. 690-1.

53. For a much detailed account see Irfan Habib, 'Inside and Outside the System: Change and Innovation in Medical and Surgical Practice in

Mughal India', in *Paper from The Aligarh Historians Society*, Indian History Congress, Millennium Session, Calcutta University, 2-4 January 2001, ed. Irfan Habib, pp. 5-8.

54. James, op. cit., p. 7.
55. Cited in James, op. cit., pp. 7-8.
56. Ibid., p. 10
57. Ibid.
58. Ibid., p. 11.
59. *Report of the Smallpox Commissioners Appointed by Government with an Appendix, Calcutta, 1st July 1850*, Calcutta, 1850, p. 36, NAI, New Delhi.
60. Ibid.
61. Ibid., p. 36.
62. *Report of the Smallpox Commissioners Appointed by Government for the Purpose of Enquiring by What Means the Extension of Smallpox can be Prevented or Rendered Less Destructive*, Calcutta, 1850.
63. Ibid., *with an Appendix, Calcutta, 1st July 1850*, Calcutta, 1850, p. 35. NAI, New Delhi.
64. Ibid., *1850*, pp. XXXII, 32-3.
65. Susan Wadley, 'Sitala, the Cool One', *Asian Folklore Studies*, vol. 39, 1, p. 45.
66. Tony Stewart, *Encountering the Smallpox Goddess: The Auspicious Song of Sitala*, Princeton: Princeton University Press, 1995, p. 390.
67. Duncan Stewart, *Report on Smallpox in Calcutta, 1833-34, 1837-38, 1843-44 and Vaccination in Bengal from 1827 to 1844*, Calcutta, 1844. V/27/857/2, OIOC, BL, London, p. 48.
68. Ibid., pp. 49-50.
69. Lalmohan Ghosal, 'Sankramak Roge Sadharaner Kartabya', *Swasthya Samachar*, vol. 1, no. 1, 1912, p. 141.
70. Ibid., pp. 50-1.
71. Ibid., p. 54.
72. Ira Klein, 'Medicine and Culture in British India', in Abhijit Dutta et al. eds., op. cit., p. 89.
73. *Report of the Smallpox Commissioners, 1850*, p. 5.
74. Ibid., pp. 5-6.
75. Ibid., p. 29.
76. Ibid., p. 30.
77. Ibid., p. 31.
78. Ibid.
79. Ibid. Appendix, p. xxii.
80. Dominik Wujastyk, 'A Pious Fraud': The Indian Claims for Pre-Jennerian Smallpox Vaccination', in G.J. Meulenbeld and Dominik Wujastyk, eds.

Studies on Indian Medical History, Delhi: Motilal Banarsidass, 2001, pp. 121-54.

81. *British Medical Journal,* 15 April 1905, p. 839. Also see John Baron, *The Life of Edward Jenner,* London: Henry Colburn, 1827, pp. 543-59.

82. *Asiatic Journal and Monthly Register,* vol. VIII, 1819, pp. 27-8.

83. *Asiatic Annual Register,* 1805, p. 76.

84. Ramnarayan Vidyaratna Bhattacharyya, *Gobij Proyog* , Calcutta: no pub., 1857.

85. Haradhan Vidyaratna Kaviraja, *Basanta Roger Nidan o Chikitsa,* Calcutta: no pub., 1868.

86. W.H. Gregg, *First Triennial Report of the Sanitary Commissioner for Bengal on the Working of the Vaccination Department in Bengal during the Three Years 1887-88, 1888-89, and 1889-90,* Calcutta, 1890, p. 9. NAI, New Delhi.

87. Ibid., p. 9.

88. From A. Halliday, Superintending Surgeon, Benaras, to Duncan Stewart, 6 November 1841. Vide Stewart's report, op. cit., p. iii.

89. *Chikitsak-o-Samalochak* ,Phalgun (February-March), BS 1301 (1894).

90. Ibid., cited in Pradip Kumar Bose ed. *Health and Society in Bengal: A Selection from Late 19th Century Bengali Periodicals,* New Delhi: Sage, 2006, p. 245.

91. From W. Findon, Superintending Surgeon, Barrackpur, 20 November 1841. Vide Stewart's report, op. cit., p. vi.

92. Reply of Baboo Ramchunder Mullick, Chore Bagan, Calcutta. *Report of the Smallpox Commissioners, Appointed by Government, with an Appendix, Calcutta 1st July 1850,* Calcutta, 1850, Appendix, p. xxii

93. Manik Bandyopadhyay, *Putulnacher Itikatha.* www.banglabookpdf. blogspot.com

94. Gregg, op. cit., p. 9.

95. H.J. Dyson, *Second Triennial Report of the Sanitary Commissioner for Bengal on the Working of the Vaccination Department in Bengal during the Three Years 1890-91, 1891-92 and 1892-93,* Calcutta: Bengal Secretariat Press, 1893, p. 16, NAI. New Delhi.

96. 'Vaccination among the Ferazi Mussalmans of Eastern Bengal', A Note by Ben H. Deare, Offg Dy Sanitary Commissioner, North Bengal Circle, in H.J. Dyson, *Triennial Report of Vaccination in Bengal during the Years 1893-96,* Calcutta, 1896, pp. xxiv-xxv Appendix. NAI, New Delhi.

97. Ibid.

98. Major H.J. Dyson, *Fifth Triennial Report of Vaccination in Bengal during the Years 1899-1900, 1900-01, and 1901-02,* Calcutta: Bengal Secretariat Press, 1902, p. 9. NAI, New Delhi.

99. Niels Brimnes, 'Variolation, Vaccination and Popular Resistance in Early Colonial South India' *Medical History*, 1 April 2004, 48(2).

100. F.C. Clarkson, *Seventh Triennial Report of Vaccination in Bengal for the Years 1905-06, 1906-07 and 1907-08,* Calcutta, 1908, p. 2. NAI, New Delhi.

101. Evidence of Pundit Moodusooden Gupta, *Report of the Smallpox Commissioners, Appointed by Government with an Appendix, Calcutta 1st July 1850*, Calcutta, 1850, p. xxxiv Appendix. NAI, New Delhi.

102. Ibid.

CHAPTER 4

Plague

This chapter engages in two major themes. First, it will explore the different perceptions of people to the epidemic plague that ravaged eastern India during the closing years of the nineteenth century. It will also describe interaction between people and the state in the struggle with the epidemc, a tragic saga that merits fresh historical interrogation. And, second, as a corollary to this, it will explore the multiple ways in which the rhetoric of Western medical intervention was contested by a popular construction of the disease. The primary objective here will be to capture the reality of the epidemic in its social setting. It will view the disease and its proposed remedial measures, the prophylactics, as functions of power and knowledge informed by the relationship between the ruler and the ruled in a colonial society. The colonial administration did not have a homogeneous perception about how to deal with the plague, and reacted differently in different situations. Nor did all categories of people react to the epidemic in a similar way; their opposition to the colonial state was not uniform either. The study seeks to interrogate how these multiple layers of perception were imbricated in the societal situation.

Studies done so far on epidemic plague in India generally concentrate on Bombay, and scholars emphasize with vengeance the plague panic and the obtrusive nature of state intervention. To David Arnold, for instance, the upsurge of public resistance to the state's intervention in Bombay was due to cultural differences and its repugnance of Western medicine.[1] To Raj Chandavarkar, on the other hand, the widespread hostility was due to the most coercive manifestation of a brutally intrusive state.[2] To Ira Klein, again, it was a conflict between Western anti-plague measures and popular culture.[3] But what about eastern India, which also shared with Bombay the phenomena of opposition to Western medicine and preventive state measures, widespread rumours, panic and scare, riots and strikes? Did it attract a similar kind of government intervention and the same form of popular resistance? Were the rumours generated during the epidemic purely 'elite discourse'?[4] Was the popular construction of the disease based only on rumour and religion rather than fact and reason? Towards the end of the century, Western medical practitioners knew very little about the disease; nor did they have any specific remedy at their disposal. In the face of the ignorance of the *firangi* doctors, what did the indigenous medical practitioners have at their disposal to prescribe for plague? What did the people in general do to circumvent the epidemic?

Evidence from eastern India, especially Bengal, indicates that the epidemic plague spread consternation throughout the province not because people were frightened by the toll of death, but especially because in the hour of crisis, the government behaved in a way that was marked by proactive haste and brute indecision. Segregation camps, inoculation, the ambulance vans, ward hospitals, plague precautions and many more measures the government resorted to were arguably the right action in the wrong place. Whatever the government did, it generated adverse public reaction, bred suspicion, and triggered rumours presumably because people were not adequately briefed about the benefits they might derive from the preventive measures. Nor had they really experienced any appreciable benefits from the measures in the past when vaccination was enforced to mitigate smallpox. Government measures might have produced the desired results had they amounted to an approximation of the traditional therapeutics and the rhetoric of Western medical intervention. People hardly found in the remedial measures anything acceptable by their collective wisdom; preventive measures were all but indigeneous.

What the people did attempt when hit by an epidemic was to turn to the gods. The urban literati suggested that a day of general prayer be observed in all churches, chapels, mosques, Brahmo Samaj venues, Hindu temples, and all places of public worship.[5] When an epidemic of cholera or small-pox had appeared earlier, *Sankirtan* parties used to be formed in different villages and towns. They chanted hymns, and went from door to door in groups, singing the name of *Hari* (God) in loud chorus, so as to drive away the epidemic. People wondered why similar parties should not be formed at the time when the fear of plague was driving them mad. Nor did they see any harm if those who were religiously inclined were to start every evening from their homes, and sing the name of the Almighty, which they believed, was sufficient to scare away the disease.[6]

THE GREAT PANDEMICS OF PLAGUE

A passage in the sixth chapter of the First Book of Samuel has been read as the earliest reference to plague.[7] The prophet says, 'one plague was on you all and on your lands. Wherefore ye shall make images of your emerods, and images of your mice that mar the land.' It is assumed that the 'emerods' (hemorrhoids) were the buboes or swellings of the groin, which are characteristic of plague, and that the 'mice' may represent recognition that plague in man was associated with rodents.

The first historically established pandemic of plague occurred in the sixth century AD and was known as the plague of Justinian. The disease broke out in Egypt and spread 'to the ends of the inhabitable world'. The number of victims of this pandemic, which lasted for fifty to sixty years, is sometimes estimated as high as one hundred million, and the historian Gibbon in his masterly summary considers this staggering figure within the bounds of possibility.

The second wave began in the neighbourhood of Constantinople and spread Westward through Europe in early 1348. It is clearly described by Guy de Chauliac, body physician to Pope Clement VI, and was the subject of nearly three hundred 'plague tracts', popular treatises which were intended to teach the people how to cope with the appalling menace. These tracts represent the first large-scale effort ever made to educate the lay public in health matters.[8] The social dislocation, which resulted from the Black Death was in some ways even more serious than the immediate mortality. A.L. Maycock has said that 'the year 1348 marks the nearest approach to a definite break in the continuity of history which has ever occurred'.

After the initial catastrophe of 1348, plague continued to rage in Europe for more than three centuries. The last major outbreak of the pandemic occurred as 'the great plague' of London in 1665, immortalized by Daniel Defoe. In the week of 19-26 September the statistical report of the city showed 8,297 deaths from all causes, of which 7,165 deaths were due to plague.[9] At the end of the seventeenth century, the disease died out in Europe; the reason for this decline remains a challenging puzzle for the epidemiologists. Many theories have been advanced such as decrease in trade with heavily infected Eastern countries, control of rats by better sanitation, and replacement of the house rat by the sewer rat, which has less intensive contact with human beings; but none of these hypotheses is wholly convincing. It is suggested by some scholars that cyclical change might occur in the virulence or in some other characteristics of the germ itself.

Plague did certainly persist in Asia, and in the late nineteenth century it broke out in the third great pandemic of this disease. Its rise began to be noticed in South China about 1870 and at Canton and Hong Kong in 1894, and shortly reached most of the larger seaports of the world. Plague appeared in Bombay some time around 1896. In fact, between 1836 and 1896, Bombay Presidency experienced the plague three times, Nagpur and Rajasthan twice, and Kumaon and Garhwal at least eight times.[10] By 1904, there were more than a million deaths from the disease in India; by the end of 1936, forty years of plague in India had killed more than 12 million people.[11]

RAT, FLEA, AND MAN

At first any contagious and fatal epidemic disease was called a plague (from *plaga*, a stroke), but this term is now applied to that particular kind of plague, which is characterized by the appearance of high fever with inflammation of lymphatic glands or bubo, and is, therefore, called the Bubonic Plague. In Sanskrit medical treatise, it has been described as *Vidradhi* and *Visharpa*, and in Yunani as *Taoon*.[12]

By the end of the nineteenth century, medical tracts emerging from Bombay sought to ascribe the disease to climate and social position. The climate and season, they argued, had 'special influence' upon the onset of plague.[13] A moderate amount of heat as a rule was found to be favourable to its occurrence, whereas a very high or low temperature was unfavourable to it. But, they argued, there had been exceptions; it had prevailed during the severest cold of winter on the Volga and in Moscow, and in extreme heat, as in Smyrna.[14] As regards social position, it was argued that poverty had been found to be the chief predisposing cause. The Great Plague of London of 1665, which was called the *poor's plague,* and of Baghdad in Mesopotamia in 1878 where it was called by *Cabiadis miseriae morbus,* are illustrations of this fact. But in Bombay, it was argued, poverty had had nothing whatever to do with the outbreak of the disease. There had been neither famine nor poverty. It broke out there among the well to do classes as well.[15]

A disease so fatal in its nature and against which human power was so helpless was in ancient times naturally attributed to the wrath of the gods. Supernatural, astrological and in some instances rational-istic causes were assigned to it. In the fourteenth century, the College of Physicians of Paris ascribed it to the influence of constellations in India. But more natural explanations followed. Putrefaction of dead animals was a cause in Egypt. Poisoning of water supply was also believed to be the cause. Undue heat, rain, watery grain, and absence of the Etesian winds were thought to generate plague.

Leaving aside the ancient theories on the causation of the disease, we find that scientists in the late nineteenth century were divided in two competing schools: first, those who believed in the germ theory and attributed the plague to specific germs, holding that germs can never arise de novo; second, those who believed that atmospheric changes and insanitary surroundings engender the seeds of pestilence which are carried through air, water or other media.[16] The arguments in favour of the first theory were, however, so strong that it was then almost universally accepted that plague was due to a specific poison which grew and multiplied under certain conditions; wherever it occurred it was caused by the implantation of those germs in a suitable soil. The soil best suited for the plague seed was one where insanitary conditions prevailed. Dirt and filth, bad ventilation, and overcrowding were supposed to be its manure.

The causative agent of plague, a bacterium, was discovered very early in the late pandemic, and the intimate relation among rat plague, the flea, and human plague was also found. During the Hong Kong epidemic the great Japanese bacteriologist S. Kitasato who had worked with Koch in Germany described a bacillus in plague-stricken patients. He showed by experiments that those bacilli if injected into lower animals produced in them symptoms of plague. A. Yersin simultaneously discovered the same germs in connection with plague. They argued that these germs must be considered to be the specific poison which produced the disease. The bacilli were found in the blood, in the buboes, and in all internal organs of the victims of the plague.[17] Thus in 1894, at Hong Kong, A. Yersin and S. Kitasato described the plague bacillus. In 1897, M. Ogata isolated the organism from rat fleas. In 1898, P. L. Simond claimed to demonstrate flea transmission.[18]

By the closing years of the nineteenth century, it was maintained that plague was a miasmatic or soil-bred disease, and that the germs found in earth, water or in some form of fermenting or decomposing material a suitable nidus for growth. In this sense, it was like malaria, which was endemic in a particular suitable area appearing and disappearing according as climate or other conditions were favourable or unfavourable. The Chinese had had a peculiar notion about the infection. They considered that the plague used to rise from the soil, and believed that it first attacked small animals with breathing organs near the soil, such as rats, then animals with breathing organs a little higher, such as poultry, pigs, dogs, goats, cows, so on till it reached man whose breathing organs are the highest.[19]

Rats are undoubtedly attacked by plague before and during its prevalence among human beings, and they play an important part in its spread. In Kumaon, this rat plague was observed by the people, and was recognized as a forerunner of the plague. *The Times of India* observed on 30 September 1896: 'It was known more than a month ago to all the people of Mandavi and to all municipal sweepers in the district that the rats were dying in thousands all over the districts. They were found dead and dying almost everywhere, and in places where dead rats were never found before.'

In such an eventuality, the people of Kumaon and Garhwal, apprehending the onset of plague, used to abandon their homestead forthwith and construct a fresh new dwelling house a little far from the epidemic zone.[20] They used to destroy their old and used clothing and wear new clothes and consequently they were spared the affliction.

It was ten years later, however, that the whole problem was analysed in a complete and strictly scientific manner by a Plague Commission appointed by the Government of India. The report of this Commission presents first a survey in various areas of cases of human plague in relation to the prevalence of the disease in rats as determined by the trapping and examination of many thousands of rats in the same localities. This study showed a close relation between the epizootic among rats and the epidemic among human beings, the course of the former preceding that of the latter by one or two weeks.

Later sections of the report give evidence to show that the direct transmission of plague from man to man is not so important a factor; insanitary conditions are significant only in so far as they favour breeding and harbouring of rats; the spread of the plague from one locality to another is generally due to imported rat fleas on the bodies or in the luggage of persons who themselves are often in good health. It is probable that the infected rat commonly plays a major role in place-to-place dissemination. In transport of the disease by sea, the rat is undoubtedly the chief agent. So far as the rat-flea-man links in the chain are concerned the 1908 study of the Indian Commission laid a sound basis of fact.[21]

PLAGUE IN EASTERN INDIA

A brief history of the disease in eastern India in its various stages of dissemination is now in order. It has been observed by a physician in a Bengali periodical that during 1836 and 1896 epidemic plague visited Bombay three times, Nagpur and Rajasthan twice, but Kumaun and Garhwal at least eight times.[22] In September 1896, almost coincidently with the outbreak of the disease in Bombay, a certain number of cases occurred in Calcutta, regarded as suspicious. These, however, after considerable discussion, were declared by the Medical Board appointed by the Government, not to be cases of genuine plague.

The first known case of plague in Calcutta was discovered in April 1898, and at the end of that month the city was declared infected.[23]

It continued so until October of the same year, when cases ceased to be reported. During these six months the attacks and deaths reported were 230 and 192 respectively. The outbreak therefore cannot be said to have been an epidemic, as that word is commonly understood.

During the next four months the disease remained quiescent. Some few isolated cases were reported which bacteriological examination showed to be plague, and there were a considerable number of cases in which the cause of death not being sufficiently apparent, it was considered advisable to disinfect the affected houses. The mortality during this period was above average, and in the opinion of the Health Officer the greater part of the excess was in fact due to plague.

It was, however, not until the end of February 1899 that the existence of plague was sufficiently demonstrated to warrant a declaration that Calcutta was again infected. This outbreak lasted until the end of June 1899, and during these four months, 1,336 cases and 1,223 deaths from plague were reported.[24] It was, however, the decided opinion of both medical officers who dealt with this period that the above figures by no means represented the actual mortality from plague, and in this opinion the Chairman concurred. What the true plague mortality was can only be conjectured. There appears to have been extensive concealment of cases, and the estimates of the authority vary, the largest being slightly over 3,000. The highest mortality was reached in March and April. From April onwards it fell rapidly, and in June the total mortality was again normal. During the second half of 1899, 1,131 cases were returned as plague, and 898 as 'suspicious'.

In 1900, the disease assumed very different proportions, and became for the first time a violent epidemic. From January to the end of June, 7,897 cases and 7,373 deaths from plague were reported.[25] Of these, 3,667 cases or nearly half of the total occurred in the month of March, and 1,996 in April, after which the virulence of the epidemic abated rapidly. During this period, cases were freely reported, and with an interesting set of figures the Chairman showed that the information obtained about deaths due to plague was complete and correct.

The total mortality due to plague in Calcutta from its first outbreak to 30 June 1900, a period of two years and three months, had been according to the official returns, 10,997. If, in accordance with the opinion of the Chairman and Medical Authorities, suspicious cases and excess mortality, not otherwise accounted for, were put down to plague, the total might be fixed at 13,000 in round numbers.

INCIDENCE BY LOCALITY

The striking feature of plague in Calcutta was the continued prominence of Ward No. V, Jorabagan, and this was still more noticeable when the figures for the latest period, which accounted for more than all the other periods together were examined. During that period no less than 1,555 cases occurred in Ward No. V or more than the total of any other two wards together. Writing of the southern portion of the ward, Dr. Simpson writes:

Everyone I have taken over the area ... is unanimous in condemning it as unfit for human habitation and a source of danger to the town.... The narrow, ill-ventilated streets; the passages to which neither light nor fresh air have access; the filthy condition of both; the close proximity of the houses to one another, and their overcrowded state, combine to form conditions which render proper sanitation impossible. It is a standing menace to the rest of the city; and should plague once obtain a firm footing in this quarter, which is the worst I am acquainted with in any city I have seen, there is every likelihood of the disease becoming endemic.

Did the epidemic hit any particular caste, community or race in Calcutta? This question was asked by several vernacular journals sometime around 1898. *Swasthya* contended that during the Hong Kong epidemic in 1894 the mortality was 18 to 19 per cent among the Europeans while 95 per cent of the indigenous Chinese fell victim to it. This, *Swasthya* argued, was not due to any special predilection of the disease to any particular caste or race; the fact that the Europeans took prompt precautions explained their relative immunity. But the Chinese evaded treatment in every conceivable way. In Calcutta too one could notice a similar situation.[26] Concealment of the disease by the indigenous population and the consequent unattended treatment explained their larger number of infection and mortality.

PROPHYLACTIC INTERVENTION REVISITED

Plague in India was primarily a disease of the long-tailed house rat, *rattus*. Other rodents were of comparatively little importance. They were less numerous, they tended to live in sewers or fields rather than in houses, and some varieties did not harbour many plague fleas. In the vast majority of cases, man was infected more or less accidentally by hungry infected rat-fleas, which had strayed from the carcasses of house-rats that had died of plague.[27] When there was epidemic plague among rats the severity of the human epidemic depended on the closeness of the association of rats and man. Human plague would cease to be a serious problem, it was believed, when we no longer provided food and shelter for rats in our houses. The real anti-plague measure, it was argued, was, therefore, 'rat-exclusion'.[28] This entailed a radical change in the habits and housing conditions of the people, which it was hoped, would come about eventually through education and social uplift.

It had been the practice all over the world to deal with plague outbreaks by certain time-honoured temporary measures, namely (i) evacuation, (ii) inoculation with anti-plague vaccine, and (iii) rat destruction by trapping and poison-baiting. These measures had undoubtedly controlled plague epidemics to a considerable extent.

When plague attacked Calcutta, Mungher, and Patna in 1898, certain measures had been prescribed by the competent authority for the whole country. The leading features were carefully enforced death registration, house-to-house visitation, segregation of the sick, the evacuation of infected localities, disinfection, and the enforcement of general sanitary precautions. The Government of Bengal too issued certain regulations to combat the epidemic.[29]

Right of Entry and Examination: The owner and occupier of any house should permit the Health Officer to enter his premises and examine any person whom he had reason to believe was infected with bubonic plague. If the person be a female who according to the custom of the country did not appear in public, the examination should be made by a woman doctor or hospital assistant.[30]

Segregation: If on examination of any person the Health Officer suspected that such a person was suffering from or infected with plague, he might cause such a person to be removed to a hospital, and might arrange for his detention, diet and treatment there. He might also cause the other occupants of the house in which such a person resided to be moved to a segregation camp and to be detained under observation for ten days. The relatives, friends, *hakims*, *vaidyas*, and priests of sick persons would be allowed access to them during day time, subject only to such precautions as the Health Officer might consider necessary.[31]

Disposal of the dead: The friends of the deceased would be permitted to dispose of the corpse by cremation or burial in accordance with their religious practices; but they should obey the directions of the Health Officer or other executive authority as to the time, route, and method of taking the corpse to the burial or cremation place. The corpses of Europeans or Mohammedans should be buried, if possible, at least six feet deep. The place of burial, if not an authorized cemetery, should be far from habitations so that there might be no risk of contaminating sources of water supply. In case cremation was the custom, the body should be completely burned at the usual burning ghat or other isolated locality in accordance with custom, the clothes brought in contact with the body being either burnt according to custom or disinfected.[32]

Inoculation: In 1900 the Government of Bengal had ordered District Officers to resort to Dr. Haffkine's system of anti-plague inoculation. But by 1903 it modified the order and argued that inoculation should be *encouraged* wherever possible, and that in places threatened by a visitation of plague, no efforts should be spared to *explain* to the people the advantage of becoming inoculated before the disease was among them.[33] It was at the same time strictly enjoined that compulsion on no account be resorted to, and that the choice whether or not to be inoculated must entirely be left to the people. In Patna Division, in Mungher, and probably in other districts of the Bhagalpur Division, nothing must be done for the present, for, it was argued, in the present temper of the people, the mere mention of inoculation would probably create a panic and intensify the difficulty of obtaining early information of the occurrence of plague.

PROPHYLACTIC MEASURES IN PRACTICE

The Government of Bengal had learnt by experience in Bombay. In eastern India the provisions regarding segregation and evacuation was at once modified by offering inoculation as an alternative, and by permitting isolation in private houses of the better classes. The people were up in arms in June 1900 in connection with plague regulations regarding evacuation and disinfection. This was particularly so at Bukhtiarpur, Dinapore (present Danapur) and Chakberia in Patna district, at Chaitantola (present Chatantola, in Jharkhand), in Monghyr and elsewhere.[34] The Lt. Governor, therefore, directed that local officers abstain from employing force for the purpose of carrying out evacuation and disinfection. They should encourage such measures by advice, by persuasion, and by supplying them with materials. If any person or class of persons nevertheless refused to comply, force should not be used for evacuation.[35]

Yet, the first appearance of plague was signaled by an un-precedented panic and exodus from the town, and by several disturbances. Fortunately, the outbreak was but a slight one, and it served the purpose of indicating how far the general policy till then prescribed was practicable in Calcutta and how far not. When plague reappeared in the succeeding year, the policy of the Government was stated in a resolution of which the following is an extract:

In order that effective measures might be taken to prevent the spread of infection, it was of the first importance that every case ... should be promptly brought to notice of the authorities. Foreign Governments would have just cause for complaint if the regulations in force were repugnant to the sense of the people affected as to drive them to withhold information and hide away their sick instead of bringing them forward for treatment.[36]

Compulsory removal to hospital and segregation were therefore abolished. The policy was to make every effort to obtain information and induce the people to accept the measures which experience had indicated to be of most use in preventing the spread of the disease. Of these, the principal, in fact, practically the only one generally used in Calcutta, had been disinfecting neighbourhoods.

Information of Plague Cases

Such measures as house-to-house visitation and the offer of rewards for information about cases had never been even suggested in Calcutta; nor had penalties for not volunteering information been enforced, except in occasional instances. Volunteer agencies had in many instances done most valuable work, but were not able to report all cases.

From the beginning of June 1898, however, a new and significant feature had appeared in the weekly returns. This was the close approximation of the number of cases with the number of deaths. In the 20 weekly returns from 2 June to the 13 October 1898, the numbers had been identical in 9, and in none of the remainder had the figures differed by more than 4.[37] This meant that if the figures were correct, virtually every case of plague terminated fatally. This was not the truth. Roughly speaking, the proportion of plague attacks that ended in death, was 80 per cent. If then it may be assumed that all deaths due to plague had been traced, such figures would indicate that during the period in question some 470 cases ending in recovery escaped discovery.

Meanwhile the total mortality from all cases had been largely in excess of the normal. The average number of deaths from all cases from 1 June to 13 October 1898 was 7,213, whereas the actuals had been 10,207, showing an excess of 2,994 or more than 40 per cent. When all due allowance had been made for the known prevalence of malaria and for deaths due directly to the floods of September, these figures were disquieting, suggesting that many deaths due to plague had escaped record as such. If these had been failure in the procedure of investigation, it is of significance; for the sheet anchor of the plague policy observed in Calcutta was disinfection of every room or apartment in which a case had occurred, whether fatal or not.

Disinfection

Of the recognized preventives against plague, disinfection had been practically the only one on which reliance had been placed in Calcutta. The procedure throughout had been to disinfect every room in which either a case of ascertained plague or a case suspected to be plague had occurred. Thus between January and June 1899, only 1,336 cases of ascertained plague were reported, but 4,750 premises were disinfected. And since February 1900, this policy was extended to the disinfection of whole *bustees* or streets in any part of which a case of plague had occurred. In 1900, up to 8 September, 42,314 rooms had been subjected to 'extra-infection'.[38]

In carrying out these measures, it was observed that the influence of men of local standing was enormous. In three instances local opposition was overcome by invoking the support of the landowners Raja Binoy Krishna Bahadur, Maharaja Sir Narendra Krishna, and Babu Sarat Chandra Mukherji respectively.

In many places human plague disappeared in the hot weather only to recur in the following cold season. It was then realized that the temperature and humidity in the depths of rat burrows were such that plague infection might continue among rats and fleas over months. Not only so, but it had been shown in some experiments in Madras Presidency in the 1930s that infected rat fleas might remain capable of conveying plague infection after as long as 29 days' starvation in rat burrows.[39]

By the 1930s certain methods of destroying fleas had been found to be of very little use. The disinfection of grain by exposure to the heat of the sun or in hot air chambers had been largely given up. The use of kerosene oil emulsion to spray plague-infected houses was of some value, but it could not affect fleas hidden in burrows or crevices or under rubbish.

A note on 'trapping and baiting as anti-plague measures' by Dr. G.D. Chitre was circulated in 1919, and the methods suggested by him had been widely employed. Various fumigants such as sulphur dioxide and carbon bisulphide had been used from time to time. Hydro cyanic acid gas was recognized as the fumigant of choice as far as the lethal effect on rats and fleas was concerned. By the 1930s experiments had been carried out in the Cumbum Valley in the Madras Presidency, and also in Bombay and Lahore under the auspices of the Indian Research Fund Association, the Travancore Durbar, and the Bombay and Punjab governments with the hope that HCN fumigation of rat burrows might prove an effective means of arresting plague and might even be successful in obliterating endemic plague.[40]

Serum Therapy

Serum therapy in plague was in 1897 at an interesting stage of development from the scientific point of view, but as a practical method of treatment it was an uncertain cure. Laboratory prophylactic experiments with anti-plague serum had been so successful that its use as a preventive or protective measure might perhaps ultimately be of use.[41] Even with a weak anti-toxic serum good results had been obtained. But as to its use as a curative agent in practical medicine, it had been a comparative failure both in Bombay and in China, in spite of statements to the contrary. This failure was scarcely to be wondered at because the serum employed was admittedly of weak anti-toxic power. Moreover, the practical difficulty of bringing patients in at an early stage for treatment meant limited success.[42]

Inoculation

Inoculation of vaccine material was carried on by Haffkine in Bombay Presidency, but with what success, it is difficult to say. Figures given of what occurred in the House of Correction, tend to show that its use was beneficial, but such figures are very incomplete. James A. Lowson, Plague Commissioner, argued in 1897 that this form of protection would never come up to that provided by an anti-toxic serum. The pain and discomfort of the inoculation was so great that unless the procedure could be 'sprung' on a whole community all at once, the results of inoculating a few were generally sufficient to cause the remainder to fight shy of any such vaccination. The laboratory work in connection with vaccine, according to Lowson, had so far not given very successful results, and the results on humans at least had not proved as striking as was desired.[43]

At first sight, the figures of 1907 with their improvement on the previous year, indicate that this measure was at last commencing to emerge from the shadow of popular disfavour. It was argued that an increase from 101 inoculations to 3,122 indicated that some of the educated at last were beginning to realize the benefits of plague prophylaxis and its absolute safety.[44] It is, however, to be noted that a few Europeans and one and two educated Bengalis had had themselves and their children inoculated; but with these extremely rare exceptions the whole of the inoculations were done on people attracted by the reward of 8 *annas*: beggars, drunkards, victims of the opium, hasheesh or cocaine habit, the drugs of the streets. Moreover, the same persons used to present themselves for inoculation two or three times within the same month. Many of them probably did a round of the different districts and were inoculated several times.[45] It was impossible to verify this from the registers, as nearly all the names in them were false.

Rat Extermination

In 1907 as many as 1,22,090 rats were killed.[46] The number seems large, but one might doubt whether it affected the rat population in the city, particularly when Kitisato reported that in Tokyo, after the destruction of 40,00,000 in three years, the result had been disappointing. In northern Calcutta, the number of rat killings was greater because the chief grain centres were situated there. Many of the grain men or *dalwallahs* had found rat collecting sufficiently paying to take up as a regular calling.[47] In other parts of the city, the rats were too scattered to make it worthwhile. In the European parts of the town, where active measures were taken by private individuals and carefully kept up, there was some diminution in the rat population.

The result of rat trapping was more educative than practical. Rewards for live rat brought in were lowered from 2 *annas* to 6 *pice* from January 1907. The attention of people was directed to the importance of rat destruction, and perhaps in the future, it was anticipated, a time would come when each householder would look after the rats of his own house. In fact, it was suggested in 1906 that the first step would be to insist that all grain godowns be rat-proof. It was argued that it should be possible to legislate for Calcutta in such a way as to cause little, if any, hardship; the law could be framed so as not to include the petty traders. As a matter of fact, some of the most recently erected dal godowns had been built pukka and rat-proof.

INTERROGATING THE MEDICAL INTERVENTION

The treatment of this pestilence might be said to have been all along empirical. The great number of medical men who had since the beginning of the nineteenth century visited the East and practised for many years in countries where plague was most prevalent had thrown no light upon its cure.[48] Leibermeister argued that the treatment of plague consisted principally in prophylaxis. The treatment of individual cases could only be anticipatory and symptomatic.[49] Aitken regretted in 1880 that recent experience had not in any degree advanced the successful treatment of the plague.[50] Dr. J.F. Payne, who had personal experience of the disease, wrote in 1885: 'No special line of treatment has proved efficacious in checking the disease once established'.[51] Dr. B. Scheube, Government Physician and Sanitary Councillor at Greiz, formerly Professor at the Medical College at Kyoto, Japan admitted that the treatment for plague was based on symptoms; unfortunately, he argued, it held out little prospect of success.[52]

To multiply the testimony of the other eminent members of the old school to the same effect would be pointless. There could not be the slightest doubt that the disease by its very nature was of the most serious kind, and attended with heavy mortality. But whether the treatment adopted by the old school, especially during the sixteenth century when Gaur was devastated by an outbreak of plague, did aggravate the disease and add to the mortality that would have resulted from it was a question which merits historical interrogation.

This much was certain: some members of the old school had admitted the injudiciousness of that treatment. Thus Dr. Gavin

Milroy, one of the greatest authorities on epidemic diseases, argued that there was little in medical writings at all satisfactory or encouraging in respect of the recovery of the sick. The perusal of recorded histories of cases of plague, he added, as observed in Malta in 1813 and in Egypt in 1835 left the impression that the patients would have fared better had they been treated with light nourishing food and cordials frequently administered, together with simple saline or acid medicines, and without active purgatives, blood-letting, or such energetic measures.

Dr. Dyson, Sanitary Commissioner of Bengal and member of the Plague Commission, condemned the use of antipyretics such as antifebrine and phenacetin, as they had been found to produce unfavourable results in plague cases. Plague is essentially a disease with adynamia of the greatest description as its pronounced condition, and therefore Dr. Dyson argued nothing should be done to aggravate that condition.

Dr. Mahendra Lal Sarkar, the celebrated homeopath, admitted in 1899: 'we have as yet had no testimony from members of our own school'. Indeed, homeopathy had not yet had a trial in this disease. And this for the simple reason that there had been no occasion for it, neither in Europe where plague had ceased to exist before homeopathy developed into a workable system, nor in America where the disease was happily unknown. Hence in homeopathy books with but few exceptions the disease was not mentioned at all.

Hence Drs. Marcey and Hint, in their *Homeopathic Theory and Practice of Medicine*, argued: 'If we may be allowed to judge of its nature from those phenomena which seem to be characteristic, we suppose the following remedies will correspond to its manifestation, and prove to it homeopathic, namely: Arsenic, Acid-nitr., Rhus-tox, Veratrum, Merc. Bell., Ipecac, Carbo-veg.' Dr. Richard Hughes in his admirable title *Manual of Therapeutics According to the Method of Hahnemann*, speaking of the Plague, says, 'Homeopathy has no practical knowledge of its therapeutic, and happily none of us are likely to have any occasion to treat it. If we had, Arsenic and Lachesis are the two medicines on which I should feel disposed to rely.'[53]

Dr. Constantine Hering stated in 1879 that Lorbacher had proposed Lachesis, Arsenic, Carbo-veg, Phosphorus and Secale as the main remedies for plague.[54] To this he added Badiaga, which he argued might be considered a remedy against the plague. To him, what Lachesis would do was uncertain. Still more uncertain was Arsenic. Dr. Hering supported Dr. Raue in administering Kali Phos, which he considered very promising. Stramonium, he argued, had more similarity to the plague symptoms than Belladonna, and Silica more than Hepar. Loimine, a preparation of the pus of the plague, brought to India by Dr. Theuille, had allegedly cured cases of the greatest importance.

Dr. Raue, who was mentioned by Dr. Hering as having proposed Kali Phos as remedy contented himself with giving the quotation from Hering under the article 'Plague' in his *Special Pathology and Therapeutics* in its 1895 edition without making any additional suggestions of his own. Dr. Winterburn, in his article on Plague in Arndt's *System of Medicine*, says

In addition to the remedies mentioned by Hering, I would like to suggest Crotalus, as possessing analogically a more intimate relation to the plague than Lachesis, Arsenicum or Phosphorus; though here again we cannot not know what is the right remedy until the patient is before us, or what a remedy will do until we have clinically tested its efficacy.

Allopathic practitioners in Bengal too joined the fray regarding prevention and treatment. The editor of *Swasthya*, a noted Bengali journal, argued that there was no known method of treatment at all.[55] Those who pretended to prescribe for its remedy, it argued, were in fact cheating the patients, and the epidemic plague had indeed provided an opportunity for the pretentious physician to mint money by prescribing fake medicines.[56]

It is, however, tempting to note what sort of treatment was going on in Bombay during the same period. Plenty of fresh air and absolute rest in the horizontal position were supposed to be the first essentials in the plague hospitals in Bombay.[57] No plague patient was allowed to sit up in bed until the temperature returned to normal or sub-normal for four or five days. In the majority of cases a preliminary calomel purge was given, while from the onset almost free stimulation by alcohol and by food was supposed to be necessary. Where many patients were brought in a condition of collapse, an ounce of stimulant mixture was administered on admission as a routine procedure. Pyrexia was treated by tepid sponging, as the usual chemical antipyretics were found to produce dangerous collapse. Ice bags applied to the head were found the best remedy for delirium, although an occasional dose of bromide was given. For vomiting when persistent, nothing succeeded better than liquor morphiae internally with ice to suck and a mustard leaf to the epigastrium.

A diet with milk, rice, congree, sago, arrowroot, egg-flip, chicken or goat soup (for Mohammedans), soda water, etc., was used during the acute stage.[58] Delirious patients frequently had to be fed by enema or the nasal tube. Subsequently, as convalescence advanced, the diet was gradually increased.

Serum treatment as advised by Drs. W.M. Haffkine and Yersin was also practised. But since the publication of Drs. Dyson and Calvert's report, Dr. G.S. Thompson, Civil Surgeon of Satara, who had subjected cases under Yersin's serum to a critical analysis, came to the conclusion that the results gave 'no better hopes of recovery than ordinary empirical treatment'. The personal impression, he argued, left on his mind was that the serum treatment did not fulfil the expectations and laudations that preceded its trial in the present epidemic. Moreover, it was not sufficiently perfected in details of strength and dosage to justify its application to patients suffering from plague. He was, therefore, inclined to think that the serum treatment of the disease, although it did enjoy a temporary reputation, would not stand the test of time.

As regards Haffkine's prophylactic serum inoculation, Indian doctors too expressed almost the same opinion. Dr. Mahendra Lal Sarkar for instance argued in 1898 that it was uncertain how long the immunity would last.[59] It was uncertain whether, when the immunity had been lost, the inoculated might not become more susceptible to the disease. And it was still more uncertain whether the inoculated, while immune themselves might not be the carriers of infection to others. It arguably introduced into the system a virulent septic poison, and Dr. Haffkine thought that the degree of immunity was faithfully proportionate to the strength and dose of the prophylactic. Hence he had proposed to increase the virulence of the poison in his future operations. In other words, he sought to introduce into the healthy human organism as virulent a culture as he could possibly make in the laboratory. Dr Sarkar argued that this was far from pleasant to contemplate for those who would like to be inoculated.

Nevertheless, George Lamb, Major, IMS and Director of Pasteur Institute of India, carried out an extensive series of researches on plague, both epidemiological and experimental, which had commanded the respect and admiration of bacteriologists over the world. He was responsible for the carrying out of that detailed enquiry into the mechanism of the epidemic spread of plague in India, the results of which had been published in five reports. He initiated and throughout bore a prominent part in the long series of experiments and observation, which resulted in the conclusive proof of the transference of plague from rat to rat, and from rats to man by the agency of fleas.[60]

Indian doctors too argued in 1897 that England had saved India from sati, infanticide and the evils of human sacrifice by thugs and Khonds. 'Let the same England save India from a greater evil still, the evil of quackery and insanitation. Give India a qualified and registered medical profession, and a well managed system of sanitation with host of other blessings will spring from this source.'[61]

Indigenous practitioners of the system of Indian medicine in Bengal had diagnosed bubonic plague to be identical to what they termed as *brodhno*, that is *baghi* (a boil in the groin). As all diseases, it was argued, were ultimately caused by imbalances between *vayu, kaffa* and *pitta*, the three constitutive elements of the body, symptomatic treatment aimed at restoring the balance would cure the disease.[62] The *Chikitsak-o-Samalochak*, an important contemporary Bengali medical journal, however, felt that it was irrational to equate *baghi* with the bubonic plague. Whatever the medical practitioners might say, there was no mention of bubonic plague in their texts on ayurveda. There was indeed mention of *vayu-, kaffa-* and *pitta-*based treatments, but there were few *kavirajas* around who really knew the significance of these terms or understood the finer points of how they worked in our system. 'These days', the editor lamented, 'ignorant pundits abound'.[63]

POPULAR RESPONSE

Popular response to plague manifested in varied forms. People reacted differently at different times to the government's acts of omission and commission.

Rumours

The measures taken by the authorities were designed with the sole object of preventing the spread of plague among the people and of protecting them from its ravages. But the motives of the government were always suspect. At different times, rumours circulated that the government desired to poison the people. The object of its action was variously stated to be the reduction of the super abundant population, the spread of the disease in order to deter foreigners from invading India, the seizure of people's money and property, the propitiation of the plague demon, and the permanent consolidation of British rule in India.[64] Among other absurd rumours was one that the intention of Government was to interfere with and destroy caste and religious observances, with the ultimate design of forcing Christianity on the 'natives' of India. Again, it was rumoured that the Viceroy had met a yogi in some remote part of the Himalayas and promised he would sacrifice two lakhs of human lives to the Goddess Kali, the Mother, to save the British Empire in India.[65]

Premangkur Atarthi, a celebrated Bengali novelist, recounts one such rumour:

One day the news spread that one woman arrived at Howrah station by Bombay Mail and then moved forward to Calcutta after hiring a carriage. When she reached Harrison Road in central Calcutta the carriage driver asked her, 'Which place do you wish to go?' The woman replied, 'Don't you know who I am? I am the goddess of plague'. Then all on a sudden she disappeared.

This story spread like a wildfire and people started shouting, 'Plague, plague, plague everywhere. Better flee the city.' People believed that plague had besieged the city and the only way to saving life consisted in fleeing.[66]

There were other rumours, with much material base, for instance that three days quarantine would be enforced, and no one would be allowed to leave Calcutta forth with. Word went around that the plague workers and the European soldiers would visit every house to examine men and women, and that inflicted persons would be immediately removed to hospital.[67]

It was further rumoured that the Viceroy was to visit Calcutta on 15 May 1898, when men and women would be inoculated forcibly. The soldiers would be called out soon and inoculation enforced compulsorily with plague serum made of cow's blood. It was even rumoured that the Queen had abdicated and that Gladstone had died of plague and that the Emperor of Russia was marching fast to Calcutta.[68]

Panic

Some physicians in Calcutta believed that the people of Kumaon and Garhwal used to flee from their homestead whenever they noticed rats dying in numbers and a few of the residents falling victim to consequent plague. They would build a new homestead nearby, abandon their old clothes, and put on new ones. They considered this change of residence a routine function as their experience taught them that they were absolutely safe in their new dwelling places.[69] By contrast, the people of Bengal behaved in a very different way. 'In the year of the first outbreak of plague in Calcutta,' writes Rabindranath Tagore in one of his novels, 'people were panic-stricken and extremely anxious, not so much for the dread of the disease, but more for the coercions and persecutions that were inflicted upon them by the peons and the orderlies who went with their official badges on.'[70]

The city of Calcutta was thrown into utter confusion when rumours, real and fabricated, spread. All were afraid of being infected. Particularly concerned about the possible violation of their privacy, many fled the city. By land, water, or railways, people began to escape. For three days a continuous stream of people began to pour out of the city like an army of ants. The fare of the carriage and palanquins increased fourfold, or sometimes even eightfold. If a carriage or palanquin were not available inspite of offering impossibly high fares, people had to settle for bullock-carts, or even a scavenger's cart. Even these were not available to the less fortunate, whose women were compelled to walk along the main streets open in public view, scorched by the midday sun.[71]

The exodus was marked by a still more pathetic sight. There were such stampedes on the boats, ships, and trains that many lost their belongings, still more had their limbs broken, and innumerable people were separated from their kin. Many lost their lives in sun stroke, mothers lost their kids, and several pregnant women had to face the danger of childbirth en route. In this way, no less than one-third of the city's population deserted.[72] The predominant fear was of being examined rather than of the disease itself. They further apprehended that the plague workers would enter the houses, and examine the women of household freely, surely outraging their modesty in the process.

To some of the Calcutta medical journals, the flight was mostly unwarranted. Experiences from Bombay, *Swasthya* argued, had shown that the epidemic had not spared the rural areas too. So escape to an apparently 'safer place' was not the remedy, it was always better to remedy the physical condition of the affected place than migrate to a supposedly innocuous village.[73] Preservation of the law of health, it argued, was prevention of disease and death.

Street Riots

While sweepers, scavengers, water-carriers, porters and labourers of the city of Calcutta struck work and ceased to attend to their respective duties, ruffians and hoodlums took to looting and instigating riots. They killed several vaccinators, beat many innocent gentlemen mercilessly and even murdered them, taking them for vaccinators. Clerks were prevented from attending office; those who disobeyed had their heads smashed. They broke tram-cars, burnt down carts carrying plague patients and even tried to damage and burn down hospitals. It was as if a torrent of unrest had overrun the city.[74]

People considered their *ijjat* to be threatened when they were subjected to corporeal inspection. The examination of railways passengers travelling to Calcutta at Khana junction had evoked vehement protest. Passengers were ordered to alight and wait in a queue after being segregated by sex. They had been publicly inspected by a European doctor. [75] It was humiliating to them and to the *ijjat* of the female folk. It was only after considerable protest that curtains were provided for the women and female doctors were employed to conduct the examination. Sarat Chandra Chattopadhyay narrates in his novel *Srikanta* how Srikanta and other passengers, males as well as females on board on their way to Rangoon were subjected to humiliating medical inspection. So callous and indifferent was the doctor in his examination of even the privates part of passengers that Sarat Chandra thought even a stuffed doll would have cried out in protest. [76]

Segregation Scare

It was generally believed that anyone infected by the plague or was suspected of carrying the germs would be summarily removed to hospital by the plague workers. The belief that during his illness or even on his deathbed, he would not be able to see his family, children, or relatives and to be comforted by them was disturbing. And 'how unbearably painful it would be if someone were unable to express his last wishes to his own people in the last moments!' [77] To an Indian, *Swasthya* argued, it is heart-rending to see a wife taken away from her husband, or a suckling baby from the lap of the mother while the infant is still alive. [78]

Ever since the plague appeared in Calcutta there had been considerable debate and agitation over the issue of isolation of patients. People were strongly opposed to segregation measures. The Calcutta literati was, however, aware that isolation was among the best means for containing the epidemic, but the way, it argued, the government had gone about enforcing it in Bombay without due consideration to prevailing social customs, defeated the very purpose.

Dr. Mahendra Lal Sarkar argued in 1898 that if all the *golmal* about the plague had produced any other effect, it had produced one thing which must be regretted, and that was the scare. [79] There had been a tremendous exodus from the town in consequence of the panic caused by the report that everybody suffering from the disease would be taken to the segregation camp for treatment. Crowds of men and women thronged the streets of Calcutta; the thoroughfares presented a scene that was shocking; and the railways stations were so overcrowded that many of those who had gone there were obliged to return home.[80] The cart pullers, coolies, *mehters*, and *dhangars* had struck work and gone back to their village homes. Home servants and cooks were also leaving the town for fear of being inoculated or segregated; it was becoming more difficult to get washermen to wash clothes. The mercantile community was meeting the greatest difficulty in removing their goods and working their godowns for want of sufficient coolies.[81] The landholders were suffering from the loss of their tenants who were leaving the city and the general public from want of menial servants of all classes.

The *doms* were perhaps hardest hit by the plague regulations. The provision that the clothes, contaminated by the plague deceased, was to be either burnt out or disinfected, was a direct intervention with their earnings. Hindu social customs enjoin that the clothes of the deceased would neither be burnt nor destroyed, but be assigned to the *doms*. But the plague regulations now deprived them of such rights. Worst still, the *doms* now interpreted it as an encroachment on their rightful claim. For the *doms* generally sold the deceased's clothes and bedding to the market to make money.[82]

Although the inoculation-scare was gradually dying out, the segregation scare was still increasing. Under the rules of the Venice Convention, the government was bound to observe them to prevent the spread of the plague. The Venice Convention held that segregation was one of the best means to prevent an epidemic of plague. But the question is whether the Convention meant that by segregation the patient was to be separated from the other members of the family and placed in a separate apartment of the house occupied by them, or that he was to be taken for treatment to a plague hospital. That the latter was not the intention of the Convention seems to be evident from the concession already made by the government that the members of a family attacked by plague, might be lodged in a separate apartment of a house, subject to the approval of the authorities. People wondered then 'why remove patients of the poorer classes to the Manicktollah Hospital, if they dread to be taken there? Then why not allow them to be treated in their own houses.'[83]

People argued that they should be treated in separate apartments of their own houses, preferably on the second floor, by their own doctors, where this was possible, under the supervision of the Health Officer and his Assistant.[84] But in cases where such accommodation was not possible the poor might be treated in the Ward Hospitals which Commissioners had opened in their respective wards, under instruction from the authorities. This, the people argued, would be in consonance with the feelings, habits, and customs of the people, and in no way raise unnecessary alarm in their minds regarding the intentions of the government.[85] Segregation of a patient from other members of the household in a confined room had all along been a custom among the Bengalis. But when it was attended by a strict enforcement of segregation law by the government, it was stoutly resisted.[86] People should be made to understand that plague was an infectious disease; it was for their own well being that the affected patient should be segregated from the rest of the members and be kept confined at a secluded part of their homestead. People were leaving the city in large numbers as they considered their hearths and homes as sacred, and did not like to be invaded by police officers.[87]

The *Times* too took up the subject of segregation scare. 'The Indian Press', it argued,

continues week after week protest against the stringency of oppressive measures, which could with difficulty be forced upon a European population, and which are so opposed to the most sacred traditions of Eastern life.... Europe understands this perfectly well in regard to Turkey. It realizes that to compel the Sultan to apply to his Oriental provinces the same that may with safety be applied in London or Berlin would shake the Ottoman Empire to its foundation.[88]

As regards the measures adopted in Bengal, the *Times* asked: 'How far is a Government justified in introducing measures at the instance of foreign advisers which can only be enforced at the cost of panic?'

Concealment

Some members of the Vigilance Committee suggested that they might be given the authority to enter the Bengali residences to examine inmates.[89] Many a Calcutta-based periodical observed that such excessively stringent laws had compelled many families to hide a patient in the house.[90] People were also afraid of being seized and forcibly carried off to hospital at the slightest indication of any kind of illness, even if it might have been a simple fever. This notion was highly entrenched in the minds of the people.[91]

Plague Hospitals

An alarm spread that a person suspected to be suffering from an attack of plague had been carried away to the isolation hospital at Manicktollah, Calcutta, which had no provision of separate wards for the treatment of Mohammedan and Hindu patients. To allay the fear, it was suggested that the *baithak-khanas* or ground floor rooms of the outer block of Bengali gentlemen's dwelling houses might be allowed to be used as segregation hospitals.[92] However, most people believed that a patient infected with plague was forced to take poison in the plague hospital in the guise of medicine, which brought instantaneous death.[93] People regarded plague hospitals as the 'Temple of Yama', the house of horrors, the place of killing the sinners. 'He has gone to Manicktollah' became a euphemism for 'He has gone to his last resting place'.[94]

Ambulance Vans

The ambulance van was also deaded. Rumour was afloat that it was constructed in such a fashion and its interior smeared with such deadly poison that even the healthy were converted to corpses.[95] People dreaded the ambulance van more than death. They believed that when a patient was put into it, his sufferings were immense, and the disinfectants made the van more uncomfortable.[96] The people argued that a *palki* should instead be used to carry the patient to the hospital. If *uriah* bearers could not be found to carry plague patients in *palkis*, *Romoni* or up-country bearers might be employed for the duty. *Palkis*, it was argued, which had no cloth covering inside, might be used as they could be easily disinfected.

Rat Killing

Rat destruction had often been unpopular with the people for religious reasons, and had more often than not received little or no cooperation on the part of the public. Marwaris objected vehemently to the killing of rats in north Calcutta. They dealt in food grains and venerated the rat as the *vahana* or carrier of lord Ganesha, the presiding deity of trade and commerce. It was also argued that any reduction of the rat population by trapping could be only temporary because rats bred so rapidly that they soon replaced the casualties. A still more serious objection to rat destruction by trapping and baiting was that the infected fleas were not destroyed. To kill rats when plague was raging merely released plague-infected fleas and increased the chances of people being infected.

Samkirtana

When the epidemic was at its worst, the urban literati suggested that a day of general prayer be observed in all churches, chapels, masjids, Brahmo Samaj venues, Hindu temples, and all places of public worship for prayers.[97]

Let us then make pujah to propitiate the Deity according to the customs of our various nationalities. Let the Hindu worship his idol, utter the mantras, which his Shastras have taught him, and make *homa* and observe other propitiatory rites. Let the Christian pray in his Church. Let the Brahmo pray in his Samaj. Let the Mahomedan utter his prayers in his Masjid. I have no doubt that the united prayers of His children would be heard and we would be freed from the plague by the Almighty Giver of all good and blessings.

In previous years, it was argued, when an epidemic of cholera or small pox had appeared in the city, *sankirtan* (devotional songs) parties used to chant hymns, and go from door to door in groups, singing the name of *Hari* in loud chorus with a view to driving away the epidemic.[98] They found no reason why similar parties should not be formed at the then crisis when the fear of plague was driving them mad.

Could the plague be prevented by *samkirtana*? The question was indeed raised in some quarters.[99] Any intelligent person, argued *Swasthya*, would admit that *samkirtana* could exert no influence on a deadly disease. 'Samkirtana is no medicine, *samkirtana* is no antidote either.' Still people believed that *samkirtana* did have an influence on the plague or any other epidemic. *Swasthya* sought to explain the reason for this belief:

If any epidemic makes its appearance in the country, the mind of the citizen is driven towards it. Day by day, even hour by hour, terrible, heart-rending wailings come up from roads, bridges, our own dwellings, those of our neighbors, the rich people's mansions and the poor people's cottages. Somebody is becoming fatherless, somebody else is losing an able son forever, someone considers this world a vale of tears after losing his wife full of so many noble qualities. Some woman who is single-mindedly devoted to her husband is more dead than alive at having to renounce all the happiness of the world on being afflicted by the death of her husband. After seeing all these heart-rending scenes repeatedly can the weak human heart retain its composure? Who will not be afraid, astonished and impatient if the noises like those made by men carrying the corpses *Balahari-Haribol, Ram Naam Satya* enters his ears? If within two to five days five to six people from out of 10 members of a family die, will those still remain alive in that house, be able to hope for life even for a single second? Therefore, the minds of the people become afraid, shaken, anxious and worried when any kind of epidemic breaks out. . . . In such a crisis do we have any alternative other than seeking refuge in God? For this reason, during the plague or during any fear of pestilence, people call upon God, chant *kirtans* in the name of Lord God, try to instill courage in their minds by focusing their attention on transcendental truths. For this reason, people also organize *samkirtana*.[100]

Interestingly, unlike Rajasthan, epidemic plague in Bengal did not produce any ritualization of godhead as its succour, nor did it usher in any invocation of deity for community worship. The Deshnoke temple of Karni Mata in Bikaner, Rajasthan, built sometime in the fifteenth century, is famous as a Rat Temple. This place is free of plague or any other contagious diseases. There are, it is believed, as many as 20,000 rats here, considered holy and not seen as pests. They are found everywhere; neither rats nor devotees are scared of each other. Karni Mata is believed to protect them all.[101]

Nevertheless, the epidemic in eastern India, believed *Swasthya*, taught the people several lessons.[102] First, people should take care of their own health by their own individual initiative, failing which they would repeatedly suffer from various epidemics in future. Second, the Health Authority in Calcutta should be much more prompt and active in the maintenance of health and hygiene in the city. Third, people should take initiative in establishing private hospitals as distinct from the government ones. This had become an imperative in view of the recurrent epidemic diseases which called for hospitalization. Indigenous hospitals, founded at private initiative and funded by the generosity of the indigenous wealthy, would take care of the time-honoured customs and rituals of the people even in the hospital premises. The patients would feel at home in such hospitals. People were not interested in the government hospitals particularly because their social and religious sensibilities were not taken into consideration. The objectives with which the Dufferin Hospital was founded could not be fulfilled because of this. The reason why people were reluctant to visit government hospitals, *Swasthya* believed, was not far to seek. Hospitals run by the government took no account of the patients' caste or social status. People were scared about losing their caste as they entered the government hospital. Second, Bengalis, both Hindu and Muslim, maintained the purdah system for their women. Such seclusion for their women folk could not be maintained in government hospitals.

Finally, the plague epidemic was teaching people still another lesson. Epidemic diseases attacked rich and poor alike; they spared none. The rich could recover fast presumably because of relatively easy access to health services, but still chances were there that he might succumb to such epidemic outbreak. Therefore, *Swasthya* argued, the rich should take the initiative in ameliorating the condition of the poor. This was part of their social responsibility as also the question of their own survival.[103]

Perusing contemporary literary tracts and vernacular periodicals emanating from colonial eastern India, it appears in retrospect that the incidence of plague was popularly viewed as a visitation of fate, and as such, one must submit to it with patience. People were also distrustful of the methods which Western science had pointed to as the most efficacious for the protection of public health and extirpation of epidemic disease. Added to this, both Hindus and Mohammedans viewed with the greatest dislike any intrusion into their homes, and especially any possible interference with the privacy of their women. Among Hindus, again, the caste system and its elaborate rules prevented intimate association such as feeding people in common. In the case of the city of Bombay, the greatest opposition was experienced from the Sunni Mohammedans, and especially the Konkani Sunnis. In the case of eastern India, the riots, strikes and disturbances were the actions of the marginalized social groups, both Hindu and Muslim, the municipal coolies, scavengers, sweepers, carters, butchers, durwans, menial servants, railways coolies, mill workers, etc., people who were hit the hardest by the plague regulations.

The situation is comparable with Pandharpur, Maharashtra, and Manjiri N. Kamat has done an excellent study of the 'strategies of evasion' of plague regulations by the pilgrims.[104] There the pilgrims and the local people contested the government policy of sanitary regulation in an indirect way. Their strategy was not overtly aggressive. But in eastern Bengal people responded in multiple ways. It ranged from use of evasive tactics to open defiance and riots. Where the incidence of plague had been really severe, the people had readily given information to the plague authorities and had sought their aid. They had also to some extent taken refuge from it in flight. But once the virulence of an epidemic had abated and the fear of death was no longer before their eyes, their attitude had throughout been one, of not of actual hostility, at least of entire indifference to precautionary measures and unwillingness to undergo the slightest inconvenience in order to guard against future calamity.

NOTES

1. David Arnold, 'Plague: Assault on the Body', in *Colonizing the Body: State Medicine and Epidemic Disease in Nineteenth-century India*, Berkeley, Los Angeles, London: University of California Press, 1993, pp. 200-39.

2. Rajnarayan Chandravarkar, 'Plague Panic and Epidemic Politics in India, 1896-1914', in Terence Ranger and Paul Slack, eds., *Epidemics and Ideas:*

Essays on the Historical Perception of Pestilence, Cambridge: Cambridge University Press, 1992, pp. 203-40.

3. Ira Klein, 'Plague, Policy and Popular Unrest in British India' *Modern Asian Studies*, 22, 4, (1988), p. 739.
4. Chandravarkar, op. cit., p. 223.
5. Gobin Chunder Dhur, *The Plague, Being a Reprint of Letters Published in the Indian Mirror for Allaying Popular Alarm and Conciliating the People to the Action of the Authorities*, Calcutta: Sanyal & Co., 1898, p. 5.
6. *Indian Mirror*, 30 May 1898.
7. C.E.A. Winslow, *Man and Epidemics*, Princeton: Princeton University Press, 1952, p. 197.
8. Ibid., p. 198.
9. Ibid., p. 199.
10. Dr. Debendra Nath Roy, 'Bubonic Plague', in *Bhisak Darpan*, November 1896, p. 185.
11. Winslow, op. cit., p. 200.
12. A. Mitra, *The Bubonic Plague*, Calcutta: Thacker, Spink and Co., 1897, p. 1.
13. 'Bubonic Plague in Bombay', a paper read by Dr. A.G. Viegas at the Meeting of the Grant College Medical Society on 24 November 1896. Bombay, 1896.
14. Viegas, op. cit., p. 23.
15. Ibid., p. 24.
16. Mitra, op. cit., p. 8.
17. Ibid., pp. 9-10.
18. C.E.A. Winslow, *Man and Epidemic*, Princeton: Princeton University Press, 1952, p. 200.
19. Mitra, op. cit., p. 12.
20. Roy, 'Bubonic Plague', op. cit., p. 186.
21. Winslow, op. cit., pp. 200-1.
22. Roy, 'Bubonic Plague', op. cit., p. 185.
23. Report on the Epidemic of Plague in Calcutta during 1898 and 1899 and up to June 1900, by E.N. Baker, Secretary (hereafter Secy) to the Government of Bengal (hereafter GOB). Home Department, Sanitary A, February 1901. National Archives of India (hereafter NAI), New Delhi.
24. Baker, op. cit., p. 7.
25. Ibid.
26. *Swasthya*, vol. 2, no. 2, Jyaistha, BS 1305, p. 36.
27. 'A note on hydro cyanic acid gas fumigation of rat burrows as an anti-plague measure', in *Health Bulletin,* no. 21, by Major W.J. Webster, Assist-Director, Central Research Institute, Kasauli, New Delhi, 1935, p. 1. OIOC, BL. V/25/850/94.

28. Webster, op. cit., p. 1.
29. *A Compilation of Regulations Issued by the Government of India and Local Governments in Connection with Plague*, Calcutta, 1898. V/27/856/6A, OIOC, BL, London.
30. Ibid., p. 89.
31. Ibid., p. 90.
32. Ibid., p. 91.
33. *The Bengal Plague Manual, being a Collection of the Extant Regulations and Executive Orders in Connection with Plague Issued by the Government of India and the Government of Bengal*, Calcutta, 1903, p. 68.
34. *The Bengal Plague Manual*, Calcutta, 1903, p. 30.
35. Ibid.
36. Baker, op. cit., p. 8.
37. Ibid., pp. 9-10.
38. Ibid., p. 10.
39. Webster, op. cit., p. 2.
40. Ibid., p. 3.
41. *Report on the Epidemic of Plague from 22nd February to 16th July 1897*, by James A. Lowson, Plague Commissioner at Bombay, Bombay, 1897, p. 40. OIOC, BL, London. V/27/856/13.
42. Lowson, op. cit., p. 40.
43. Ibid., p. 41.
44. *Report on Plague in Calcutta for the Year Ending 30th June 1907*, by T. Frederick Pearse, ed. W.C. Hossack with Appendices by W.C. Hossack and H.M. Crake, Calcutta, 1907. NAI, New Delhi.
45. Pearse, op. cit., p. 3.
46. Ibid.
47. Pearse, op. cit., p. 4.
48. James Copland, *Dictionary of Practical Medicine*, cited in *Therapeutics of Plague, Being Suggestions for the Prophylactic and Curative Treatment of the Disease*, by Mahendra Lal Sarkar, Calcutta, 1899, p. 2. Oriental and India Office Collection (OIOC), British Library (BL), London.
49. *Zimssen's Cyclopaedia*. Vide Mahendra Lal Sarkar, op. cit., p. 3.
50. William Aitken, *Science and Practice of Medicine*, London: R. Griffin, 1880.
51. J.F. Payne, *Encyclopaedia Britannica*, 1885, 9th edn.
52. B. Scheube, *The Diseases of Warm Countries: A Manual for Medical Men*, London: John Bale, Sons & Danielsson Ltd., 1903.
53. Sarkar, op. cit., p. 6.
54. *North American Journal of Homoepathy*, August 1879.
55. *Swasthya*, vol. 2, no. 1, Baisakh 1305 (bs), p. 11.
56. Ibid., p. 12.

57. Drs. Dyson and Calvert's Report, para 24. Cited in Mahendra Lal Sarkar op. cit., p. 45.
58. Drs. Dyson and Calvert's Report, para 25. Cited in M.L. Sarkar, op. cit., pp. 46-7.
59. *Calcutta Journal of Medicine*, June 1898.
60. Lamb Papers. MSS Eur D 893. OIOC, BL, London.
61. U.L. Desai, *Plague in India*, Simla: Government Central Printing Office, 1898, p. 8.
62. *Chikitsak-o-Samalochak*, Pous (December-January), BS 1303 (1896-7), cited in Pradip Kumar Bose, ed., *Health and Society in Bengal*, New Delhi: Sage, 2006, pp. 190-1.
63. *Chikitsak-o-Samalochak*, Pous (December-January), BS 1303 (1896-7).
64. *The Bengal Plague Manual, Being a Collection of the Extant and Executive Orders in Connection with Plague Issued by the Government of India and the Government of Bengal*, Calcutta, 1903, p. 29.
65. *Amrita Bazar Patrika*, 4 May 1898 and 9 November 1898.
66. Premangkur Atarthi, *Mahasthabir Jatakta*, pt. 1, Calcutta: Dey's Publishing, 1981, pp. 46-7.
67. *Swasthya,* Shraban (July-August), BS 1305 (1898).
68. Letter from Secretary, Government of Bengal, to the Secretary to the Government of India, Home Dept, 28 May 1898, Home (Public), June 1898, Proceedings B, nos. 91-2, NAI, New Delhi.
69. Dr. Debendranath Ray, 'Bubonic Plague', in *Bhishak Darpan*, November 1896, p. 186
70. Rabindranath Tagore, *Chaturanga*, pp. 447-8.
71. *Swasthya,* Shraban (July-August), BS 1305 (1898), tr. Pradip Kumar Bose, *Health and Society in Bengal: A Selection from Late 19th-Century Bengali Periodicals*, New Delhi: Sage, 2006, pp. 212-13.
72. Ibid. Cited in Pradip Kumar Bose, *Health and Society in Bengal*, p. 214.
73. *Swasthya,* vol. 5, no. 1, Baisakh BS 1305 (1898), p. 12.
74. *Swasthya,* Shraban (July-August), BS 1305 (1898), cited in Bose, ibid., p. 214.
75. Arnold David, 'Touching the Body: Perspectives on the Indian Plague, 1896-1900', in Ranajit Guha, ed., *Subaltern Studies*, vol. V, New Delhi: Oxford University Press, 1987, p. 64.
76. Srikanta, in *Sulabh Sarat Samagra*, vol. 1, Calcutta: Ananda Publishers, 1918, pp. 17-18.
77. *Swasthya,* Shraban (July-August), BS 1305 (1898).
78. Ibid.
79. *Calcutta Medical Journal,* May 1898.
80. *The Indian Mirror*, 5 May 1898.
81. Ibid., 11 May 1898.

82. Amal Das, 'Plague and People in Calcutta,1898-1900', in Abhijit Dutta et al. (eds.), *Explorations in History: Essays in Honour of Professor Chittabrata Palit,* Kolkata: Corpus Research Institute, 2003, pp. 127-8.

83. *The Indian Mirror,* 10 June 1898.

84. *Swasthya,* vol. 2, no. 1, Baisakh BS 1305, p. 12.

85. *The Indian Mirror,* 11 June 1898.

86. *Swasthya,* vol. 2, no. 1, Baisakh BS 1305, p. 17.

87. Ibid., pp. 16-17.

88. Cited in *The Plague,* by Gobin Chunder Dhur, Calcutta: no pub., 1898, p.18.

89. *The Statesman,* 17 June 1898.

90. *Swasthya,* Baisakh (April-May) BS 1305 (1898).

91. *Swasthya,* Shraban (July-August), BS 1305 (1898).

92. *The India Mirror,* 23 June 1898.

93. *Swasthya,* Shraban (July-August), BS 1305 (1898).

94. Nield J. Cook , 'Report on Plague in Calcutta', in *Report of the Plague in Calcutta During the Years 1898-99, 1899-1900 and upto 30th June 1900,* Calcutta, 1900, p. 15.

95. *Swasthya,* Shraban (July-August), BS 1305 (1898)

96. *The Indian Mirror,* 17 June 1898.

97. Ibid., 29 May 1898.

98. Ibid., 1 June 1898.

99. *Swasthya,* Chaitra (March-April), BS 1306 (1900).

100. Ibid., tr. Pradip Kumar Bose, *Health and Society in Bengal: A Selection from Late 19th-Century Bengali Periodicals,* New Delhi: Sage, 2006, pp. 226-8.

101. http://temple-stories.blogspot.com.Accessed on 24.08.2015. I am also indebted to Sudit Krishna Kumar, my colleague in the Department for this piece of information.

102. *Swasthya,* vol. 2, no. 2, Jyaistha, BS 1305, pp. 42-3.

103. Ibid., pp. 44-5.

104. Manjiri N. Kamat, '"The Palkhi as Plague Carrier": The Pandharpur Fair and the Sanitary Fixation of the Colonial State; British India, 1906-1916', in Mark Harrison and Biswamoy Pati, *Health, Medicine and Empire: Perspective on Colonial India,* New Delhi: Orient Longman, 2001.

CHAPTER 5

Conclusion

One can see that the four major epidemic diseases that the study interrogates could be situated at different locations with differing perspectives and varied meanings. The politics of location, as one could notice, enmeshed in an interlocking relationship, sometimes between the disease and the victim, between the patient and the doctor, between the society and the individual, and more often than not between the state and the people in a colonial territory. I have sought to see how health and illness were shaped, experienced and understood at different locations at different times in the light of historical and political forces. Understandably, our focus very often reverts to think of epidemics geographically instead of just historically. This geographical imagination of an epidemic helps us track its migratory and trans-cultural formations and make it more viewable within a temporal and spatial framework.

The major thrust of my study has been popular perception of the epidemics. Popular construction of disease incidences is indeed very difficult to document historically. Such attempts always tend to suffer from the problem of representation. Official records of the popular construction were often constrained by lack of adequate knowledge of people's life, livelihood and social customs in the countryside. Evidences of the colonial officers should, therefore, be read with caution.

Again, if archival sources are to be used with care, much greater caution should be exercised with regard to literary evidences. Contemporary literature on the subject was primarily the discourse of the urban literati. Quite a few writers were familiar with the travails of the rural poor, and empathized with victims of epidemics as such. Nevertheless, they need not necessarily always reflect the authentic perceptions of the suffering masses. Village literati might have rescued us from such constraints; literary tracts originating from the villages might have been much more dependable. But such literary tracts do not come in abundance. Under such circumstances, oral tradition might, admittedly, convey the authentic perception of the ailing masses. In this study, a few conclusions have been drawn from oral evidence collected from some octogenarian villagers.

The most ancient of all theories regarding the causation of sickness was the assumption that disease, particularly epidemic ones, was the result of the malign influence of supernatural powers.[1] These powers might be exercised by living persons (witches), or by the spirits of the disembodied, or even by a super-human entity. If the disease was supposed to be due to personalized anthropomorphic entities, one should obviously be able to influence the demonic personalities by propitiation through sacrifice, by exorcism, or frightening the evil spirit through invocation of more powerful spiritual forces.

This fatalistic disposition of disease is neither special to the orient nor alien to the Western world. In pre-modern Europe, every Christian believed that life was not a lottery, but reflected the working out of God's purposes. If things went wrong, he did not have to blame his luck but could be assured that God's hand was at work.[2] The victim of misfortune would thus draw some consolation from the belief that some suffering was essential as proof that God retained an interest in him. Temporal afflictions were interpreted as signs of God's affection.[3]

Most of those who saw the Great Plague in Europe as the product of divine wrath assumed that God worked through natural causes, bringing the epidemic by contagion or by the putrefaction of the air. The theologians, however, believed that there was little to be hoped from natural remedies until the patient himself repented his way.[4] The correct reaction on the part of a believer stricken by misfortune was, therefore, to search in him the moral defect that had provoked God's wrath. It was also customary for national disasters to be regarded as God's response to the sins of the people: poverty, death, and famine were caused by God's anger at the vices of the community.[5]

In modern times, medical scientists relate disease to mundane causes, and not any supernatural intervention. Speaking about the constant challenge to human health in the tropics, recent scholars relate it to the 'biological environment of human settlement'.[6] The mortality and morbidity in tropical countries, they argue, 'is not racial, but environmental', for whenever 'planned reforms are carried through, improvement follows'.[7]

In India, however, people in the countryside were not always amenable to reason; neither did they always go for a conclusion adhering to a logical and rational thinking. In an account of life in an Indian village in Hyderabad State, an anthropologist found illness related to the ritual structure of Hindu life.[8] Only a punctilious observance of the ritual cycle of festivals leads to the prosperity of the family. Most diseases, the anthropologist found, are interpreted as a 'fault in the physical system', and are treated with herbal medicine or modern drugs obtained from a dispensary. But persistent headache, intermittent fever, or repeated abortions are attributed to supernatural forces. Smallpox, cholera and plague were attributed to the wrath of various goddesses, and for these diseases, worship was the only remedy.[9] Another anthropologist who has worked in north India argues that many standard rites are in fact precautionary ceremonies in which a deity is regularly propitiated so that he will only do good to his worshippers.[10]

In colonial Bengal, popular reaction to epidemic from different sections of society generally offer a mixed and complex pattern. This complexity stems from widely differential perceptions of the nature of an epidemic. So far as the question of epidemic smallpox and the subsequent government attempts at vaccination was concerned, popular reaction was hostile. Not infrequently, Hindus objected to it for fear of loss of religion, and Muslims were opposed to it for fear of continued political subjugation under alien rule with increasing threat to their beliefs. Vaccination to both the communities was interpreted as imperialist design fraught with socio-political danger. Epidemic cholera called forth vehement popular criticism of the government acts of omission and commission. People sought to find remedy in propitiation of a popular deity. The Hindus worshiped Olai Chandi and the Muslims Ola Bibi, thus forging a meaningful religious syncretism in the moments of existentialist crisis. Plague evoked much greater protest against prophylactic measures in Bombay than in eastern India. So far as epidemic malaria was concerned, the response of the masses, the rural gentry, the intelligentsia and indigenous medical practitioners, was one of stoic acceptance, and not dreaded repulsion. Taken as a whole, public reaction to all the disease and the consequent government intervention never took a form of widespread and open revolt, and never went beyond the limits of control.

There is, however, nothing new in this kind of attitude, for an epidemic as such had never been a dreaded phenomenon in eastern India. It has all along been depicted in Bengali folklore as a living entity. Narrators of Bengali folk tales depict an epidemic as a cute village belle who delights in frequenting village huts, and though often despised by the wealthy landlords is in fact welcomed by the penniless peasants.[11] Bengali poets in fact took pride in the stoic indifference of the masses to epidemics and destruction.[12]

Generally speaking, reaction to the attitude of the government hardened in the course of time. It followed a more or less a uniform pattern with marginal variations. Almost all sections of society raised their accusing finger to the colonial government for bringing sudden and untimely death into their lives. The accusation was sometimes violent, sometimes muted. Reputed doctors held the rulers responsible for the epidemics, but instead of blaming the rulers, they scolded the subjects for their apathy to personal hygiene. They constructed the epidemic from a purely medical perception, and there perhaps they erred in judgement. They could have realized that the struggle against disease must begin with a war against bad government. Men will be totally cured, rightly observed Foucault, only if at first liberated. In a society that was free, and inequalities reduced, the doctors would have no more than a temporary role: that of giving legislator and citizen advice as to the regulation of his heart and body.[13]

Early colonial Bengal witnessed the function of multiple medical systems among different communities. This led to some sort of syncretism and contestation at multiple levels of society. The predominant medical practice was ayurveda. Until the early nineteenth century, the attitude of the colonial rulers was marked by tolerance.[14] At times, some British practitioners were even found to be appreciative of the Indian system of ayurveda because Western medicine and ayurveda, it was believed, shared many features. However, it has recently been argued that ayurveda and other Indian medical traditions were invented as nationalist alternatives to Western biomedicine.[15] From the colonial government's perspective, on the other hand, ayurveda was located as a fragment of 'subaltern' science vis-à-vis 'Western' science.[16] Gradually, therefore, colonialism created a situation in which the state subscribed to only one system of medicine and that became the dominant official system. Side by side, the state also advanced a discourse about its subjects who became their 'gaze' of investigation.[17] In 1822, in Calcutta, the government established the Native Medical Institution where the indigenous system of medicine was integrated with Western medicine and was taught in Sanskrit and Urdu.[18] In course of time, the indigenous system was marginalized, as a 'closed' system, opposed to Western scientific methods. The government, therefore, sought to address various epidemic diseases through Western medical intervention.

In fact, it has recently been argued very rightly that universalist claims of Pasteurian and Kochian bacteriology were starkly contradicted when introduced to post-Mutiny India.[19] The study of microbes in the tropics did never amount to any major epistemological change; rather it 'intersected with prevailing prejudices to animate dialectic between germs and climate'.[20] In imperial thought, cultures, both bacteriological and anthropological, 'behaved differently in hot climates'.[21] The British legitimized their paternalistic control over India by rendering the tropical environment safer and helped her escape from the imperial narrative of disease, filth, and putrefaction.[22]

Coming to the issue of the doctor-patient relationship, things have evolved very little. Medical paternalism has hardly transformed into patient autonomy. The medical surveillance used to operate as a structure of control in the new hospital system. The patient was shut out and isolated from his own world: community, family care and enshrouding concern for parents, wife and children. The panoptic layout of the hospital ward worked wonders to maintain strict control. Upon entering the hospital, a person became a patient, shorn of autonomy. The loss of agency might be seen in something as subtle as a change in dress from a man's own clothing to a hospital gown. Knowledge was crucial to this situation, and the scientific knowledge of treating an epidemic disease assigned the doctor an overriding primacy over the patient who was at the receiving end. The whole process reaffirmed the hierarchical relationship between patient and doctor.[23]

Thus technology-driven Western medical system under colonial dispensation tended to commercialize private lives and stripped of the patient his subject hood. The newly acquired medical knowledge as enjoined by the Western medical system had driven a wedge between the patient and the medical practitioner. This division underlined a distinction between the knowledge of the physician acquired through apprenticeship and a new formal education that came with the burgeoning medical schools across the country, and the hapless victim of an epidemic who had tried in vain all the potentials of indigenous remedial systems that tried to survive through neglect by the colonial government.

In fact, little inter-faith dialogue existed between the patient and the physician when epidemic struck terror in colonial Bengal. The patient had his own notion of the disease, informed by his certain value-laden cultural specificities of the time, the indigenous *daktar* persisted in the faith in the capacity of the patient to withstand falling ill, and the colonial doctor had his own ideas premised on an overarching colonial discourse of power/knowledge. Society, for its part, apprehended that its rigoriously idealized self-image of a sanitized cultural space was about to be subverted by the technology-driven intervention. The systematizing claims of science would circumscribe its authority during the days of epidemic malaria, small-pox, cholera and plague.

NOTES

1. C.E.A. Winslow, *Man and Epidemics*, Princeton: Princeton University Press, 1952, p. 7.
2. Keith Thomas, *Religion and the Decline of Magic*, London: Penguin Books, 1980, p. 91.
3. Ibid., pp. 93-5.
4. Ibid., p. 100.
5. Ibid., p. 96.
6. Margaret Read, *Culture, Health and Disease: Social and Cultural*

Indifferences on Health Programmes in Developing Countries, London: Tavistock Publications, 1966, p. 5.

7. Arnold Sorby, *Medicine and Mankind*, London: Faber and Faber, p. 195.

8. S.C. Dubey, *Indian Village*, London: Routledge and Kegan Paul, 1955, p. 127.

9. Ibid.

10. M.E. Opler, 'The Cultural Definition of Illness in Village India', *Human Organization*, no. 22, 1963, pp. 32-3.

11. Gaur Kishore Ghosh, *Jal Padre Pata Nadre*. Serialized in *Desh* in 1954.

12. Satyendranath Datta writes: 'We never die of famine; we in fact flourish on epidemic.' 'Amra, Kuhu Keka', *Kabita Sangraha*, Satyendranath Dutta, Kolkata: Paschimbanga Bangla Academy, 1319, p. 154.

13. Michel Foucault, *The Birth of the Clinic: An Archaeology of Medical Perfection*, London: Tavistock Publications, 1973, p. 33.

14. Mark Harrison and Biswamoy Pati, eds., *The Social History of Health and Medicine in Colonial India*, London and New York: Routledge, 2009, p. 9.

15. David Hardiman, 'The Invention of Indian Tradition', *Biblio*, 12, 2007, pp. 24-5.

16. David Arnold, *The Tropics and the Travelling Gaze: India, Landscape and Science 1800-1856*, New Delhi: Permanent Black, 2005.

17. V. Sujatha and Leena Abraham, 'Medicine State and Society', in *Economic and Political Weekly*, 18 April 2009, vol. XLIV, no. 16, p. 36.

18. Zhaleh Khaleeli, 'Harmony or Hegemony? The Rise and Fall of the Native Medical Institution, Calcutta in 1822-35', *South Asia Research*, 21, 2001, pp. 77-104.

19. Pratik Chakrabarti, *Bacteriology in British India: Laboratory Medicine and the Tropics,* Rochester, New York: Rochester University Press, 2012.

20. Peter Hobbins, *Health and History*, vol. 16, no. 1, 2014, p. 122.

21. Chakrabarti, op. cit., p. 10.

22. Ibid. p. 3.

23. I am particularly indebted to Dr. Jayanta Bhattacharya for this insight. For details see https://www.scribd.com/doc/23711842, accessed on 22 February 2011.

Bibliography

A. PUBLISHED PRIMARY SOURCES

WEST BENGAL STATE ARCHIVES, CALCUTTA

1. Proceedings of the Government of Bengal

Sanitation Branch:
   (i) General Department
       From April 1868 to April 1869
       June 1870 to April 1878
  (ii) Political Department
       From May 1869 to May 1870, June 1872
 (iii) Finance Department
       From May 1878 to December 1879
  (iv) Municipal Department
       From January 1880 to August 1887
       January 1889 to December 1910
       January 1914 to December 1920
   (v) Judicial Department
       From September 1887 to December 1888

2. Official Reports

   (i) *Transaction of the First Indian Medical Congress.* H.J. Dyson, Calcutta: Calendonian Steam Printing Works, 1894.
  (ii) *Studies in Malaria as it Affects Indian Railways*, pt. II, R.A. Senor-White and C.D. Newman, Calcutta: Govt. India Cent. Pub. Br., 1932.
 (iii) *Report and Methods of Malaria Control on the Eastern Bengal Railways.* R.J.L. Slade, Calcutta: Bengal Secretariat Press, 1927.
  (iv) *Report of the Malaria Survey of the Jalpaiguri Dooars.* A.D. Stewart, Calcutta, Government Press, 1926.

(v) *Report of the Drainage Committee, Bengal.* G.E. Stewart and Proctor, Calcutta: Bengal Secretariat Press, 1907.

(vi) *First Report on Malaria in Bengal.* A.B. Fry, Calcutta: Bengal Secretariat Book Depot, 1912.

(vii) *Second Report on Malaria in Bengal.* A.B. Fry, Calcutta: Bengal Secretariat Book Depot, 1914.

(viii) Statistical Department. *Prices of Food Grains, Fire Woods and Salt in Bengal for the Years 1866 to 1878*, Calcutta.

(ix) Revenue Department. *Report on the Condition of the Lower Classes of Population in Bengal.* Calcutta: Bengal Secretariat Press, 1888.

(x) *Report on the Rivers of Bengal.* W.S. Sherwill. Papers of 1856, 1857, 1858 on the Damodar Embankments.

(xi) Public Health Department. *Report on the Working of the Anti-Malaria Campaign in Rural Areas of Bengal with Quinine and Plasmochine, 1933-34*, Calcutta.

(xii) Irrigation Department. *Annual Report on Major Rivers, Bengal, 1932-33*, Calcutta.

(xiii) *Report of the Meteorological Reporter to the Government of Bengal, Meteorological Abstract Annual, 1867-68, 1968-69, 1870-72*, Calcutta.

(xiv) *Meteorological and Rainfall Table of the Province of Bengal, 1887 to 1889*, Calcutta.

(xv) *Extracts Referring to the Connection between Obstruction to Drainage and Malarious Fever, 1894*, Calcutta.

(xvi) *Report on the Control and Utilization of Rivers and Drainage for the Fertilization of Land and Mitigation of Malaria.* Sir Edward Buck, Calcutta: Bengal Secretariat Press, 1907.

(xvii) *Final Report on the Survey and Settlement Operation in the District of Jessore, 1920-24.* M.A. Momen, Calcutta: The Bengal Secretariat Book Depot, 1925.

(xviii) *First Report on the Survey and Settlement Operations in the Bakarganj District, 1900 to 1908.* J.C. Jack, Calcutta: The Bengal Secretariat Book Depot, 1915.

(xix) *Final Report on the Survey and Settlement Operations in the Faridpur District, 1904 to 1914.* J.C. Jack, Calcutta: The Bengal Secretariat Book Depot, 1916.

(xx) *Report on the Cultivation of, and Trade in Jute in Bengal and on Indian Fibres for the Manufacture of Paper.* H.C. Kerr, Calcutta: no pub., 1874.

(xxi) *Annual Sanitary Report for Bengal, 1866—with Appendices Containing Returns of Sickness and Mortality among British and Native Troops and also among the Prisoners in the Bengal Presidency.*

(xxii) *Annual Report on the Working of Hospitals and Dispensaries, 1868-69.*

(xxiiii) *Endemic Fever of Lower Bengal.* J.G. French, Calcutta: no pub., 1874.

(xxiv) *Report on Epidemic Fever in Burdwan and Birbhum during 1871.*

(xxv) *Report of the Commission Appointed in 1864 to Enquire into the Nature and Probable Causes of the Epidemic Fever in the Districts of Hooghly, Burdwan, Nadia and 24-Parganas, 1900.*

(xxvi) *Food Grain Supply and Famine Relief in Bihar and Bengal.* A.P. Macdonnel, Calcutta: no pub., 1874.

(xxvii) *Report on the Internal Trade of Bengal for the Year 1876-7.*

(xxviii) *Sanitary Commissioner's Annual Report for 1873.* C.J.J. Jackson, Calcutta, 1874.

(xxix) *Tenth Annual Report of the Sanitary Commissioner for Bengal, 1877.* Robert Harvy, Calcutta: Bengal Secretariat Press, 1879.

(xxx) *Nineteenth Annual Report of the Sanitary Commissioner for Bengal for the Year 1866.* R. Lidderdale, Calcutta: Bengal Secretariat Press, 1867 .

(xxxi) *Twenty-first Annual Report of the Sanitary Commissioner for Bengal for the Year 1888.* W.H. Gregg, Calcutta: Bengal Secretariat Press, 1889.

(xxxii) *Twenty-second Annual Report of the Sanitary Commissioner for Bengal for the Year 1889.* W.H. Gregg, Calcutta: Bengal Secretariat Press, 1890.

(xxxiii) *Twenty-third Annual Report of the Sanitary Commissioner for Bengal for the Year 1890.* W.H. Gregg, Calcutta: Bengal Secretariat Press, 1891.

(xxxiv) *Thirty-third Annual Report of the Sanitary Commissioner for Bengal for the Year 1900.* H.J. Dyson, Calcutta: Bengal Secretariat Press, 1901.

(xxxv) *Report on Malaria Control of Damodar Valley,* Calcutta: no pub., 1954.

(xxxvi) *Report on Anti-Malarial Operation at Dinajpur.* C.A. Bentley, Calcutta: Bengal Secretariat Book Depot, 1913.

NATIONAL ARCHIVES OF INDIA, NEW DELHI

1. Proceedings of the Government of India

Home Department, Sanitation Branch

2. Official Reports

A Compilation of Regulations issued by the Government of India and Local Governments in Connection with Plague, Calcutta: Office of Government Print, 1898.

Aykroyd, W.R., *The Nutritive Value of Indian Foods and the Planning of Satisfactory Diets,* Health Bulletin no. 23, Delhi: Government of India Press, 1939.

Balfour Edward, *Cholera: Are There Towns or Villages in India where Cholera has never Appeared from the Period of its Outbreak in 1817?,* Madras: Fort St. George Gazette Press, 1852.

Bannerman, W.B., *Statistics of Inoculations with Haffkine's Anti-Plague Vaccine, 1897-1900,* Bombay: Government Central Press, 1900.

Banks, Charles, *Observations on Epidemics of Cholera in India, with Special Reference to their Immediate Connection with Pilgrimages,* Cuttack: no pub., 1896.

Baker, E.N., *Report on the Epidemics of Plague in Calcutta during 1898 and 1899 and up to June 1900,* Calcutta: Bengal Secretariat Press.

Bellow, H.W., *Cholera in India 1862 to 1881: Bengal Province 1862 to 1881 and Review,* Calcutta: Bengal Secretariat Press, 1884.

Bryden, James L., *Epidemic Cholera in the Bengal Presidency,* Calcutta: Office of the Supdt. of Govt. Printing, 1869.

————, *Epidemic Cholera in the Bengal Presidency: Note on the Epidemic Connection of the Cholera of Madras and Bombay with the Cholera Epidemics of the Bengal Presidency,* Calcutta: Office of the Supdt. of Govt. Printing, 1871.

Choudhury, A.C., *Report on an Enquiry into the Standard of Living of Jute Mill Workers in Bengal,* Calcutta: Bengal Secretariat Book Depot, 1930.

Corbyn, Frederick, *A Treatise on the Epidemic Cholera, as it has Prevailed in India; Together with the Report of the Medical Officers, Made to the Medical Boards of the Presidencies of Bengal, Madras and Bombay, for the Purpose of Ascertaining a Successful Mode of Treating that Destructive Disease; And a Critical Examination of All the Works which have Hitherto Appeared on the Subject,* Calcutta: W. Thacker & Co., 1832.

Clarkson, F.C., *Seventh Triennial Report of Vaccination in Bengal for the Years 1905-1906, 1906-1907 and 1907-1908,* Calcutta: The Bengal Secretariat Book Depot, 1908.

Condon, J.K., *The Bombay Plague; Being a History of the Progress of Plague in the Bombay Presidency from September 1896 to June 1899,* Bombay: Education Society's Steam Press, 1900.

Couchman, M.E., *Account of Plague Administration in the Bombay Presidency from September 1896 till May 1897*, Bombay: Government Central Press, 1897.

Cunningham, J.M., *Report on Cholera Epidemic of 1875 in Northern India*, Calcutta: Office of the Supdt. of Govt. Printing, 1876.

————, *Cholera: What Can the State Do to Prevent it?* Calcutta: Office of the Supdt. of Govt. Printing, 1884.

Dyson, H.J., *Second Triennial Report of the Sanitary Commissioner for Bengal on the Working of the Vaccination Department in Bengal during the Three Years 1890-91, 1891-92 and 1892-93*, Calcutta: Bengal Secretariat Press, 1893.

————, *Fifth Triennial Report of Vaccination in Bengal during the Years 1899-1900, 1900-01 and 1901-02*, Calcutta: Bengal Secretariat Press, 1902.

Edwards, M.I. Neal, *Report of an Enquiry into the Causes of Maternal Mortality in Calcutta*, Health Bulletin, no. 27, New Delhi: Manager of Publications, 1940.

Gatacre, W.E., *Report on the Bubonic Plague in Bombay*, Bombay: Times of India Steam Press, 1898.

Graham, J.D., *Report on an Experiment in Quininization of Badaun District School*, 1910.

Gregg, W.H., *First Triennial Report of the Sanitary Commissioner for Bengal on the Working of the Vaccination Department in Bengal during the Three Years 1887-88, 1888-89, and 1889-90*, Calcutta: Bengal Secretariat Press, 1890.

Hamilton, Francis, *An Account of the District of Purnea in 1809-10*, Patna: Bihar and Orissa Research Society, 1928.

Hunter, G.Y., *Health in India: Medical Hints as to Who Should Go There, and How to Return Health There and on Returning Home*, Calcutta: no pub., 1873.

James, S.P., *Malaria in India,* Calcutta: Periodical Publications, 1902.

Jameson, James, *Report on the Epidemic Cholera Morbus in 1817-19*, Calcutta: Government Gazette Press, 1820.

Lowson, James A., *Report on the Epidemic of Plague from 22nd February to 16th July 1897*, Plague Commissioner, Bombay, London, 1897.

Martin, Montgomery, *The History, Antiquities, Topography and Statistics in Eastern India*, London: W.W. Allen and Co., 1838.

McCay, Capt. D., *Investigation on Bengal Jail Dietaries with Some Observations on the Influence of Dietary on the Physical Development and Well-being of the People of Bengal*, Calcutta: Supdt. Govt. Printing, 1910.

Memorandum on the Science and Art of Indian Medicine, Committee of the Indigeneous Systems of Medicine, Madras, 1948.

Murray, John, *Report on the Treatment of Epidemic Cholera*, Calcutta: Supdt. Govt. Printing, 1869.

Murti, G. Srinivasa, *Memorandum on the Science and the Art of Indian Medicine Presented to the Madras Government Committee on the Indigenous System of Medicine,* Calcutta: Bengal Secretariat Press, February 1923.

Pearse, T. Frederick, *Report on Plague in Calcutta for the Year Ending 30th June 1907*, Calcutta: Bengal Secretariat Press, 1907.

Report of the Smallpox Commissioners Appointed by Government for the Purpose of Enquiring by What Means the Extension of Smallpox can be Prevented or Rendered Less Destructive, Calcutta: Government Press, 1850.

Report of the Smallpox Commissioners Appointed by Government with an Appendix, Calcutta, 1st July 1850, Calcutta: Government Press, 1850.

Report of the Health Officer for the City of Calcutta, Calcutta, 1922.

Report on Plague in Calcutta for the Year Ending 30th June 1907, by T. Frederick Pearse, ed. W.C. Hossack with Appendices by W.C. Hossack and H.M. Crake, Calcutta: Corporation Press, 1907.

Report on the Hygienic Measures to be Adopted for Preservation against Asiatic Cholera with an Appendix on Disinfections as Applied to Cholera, Calcutta (publication date unavailable), International Sanitary Conference.

Roberts, E., *On Some Practical Methods of Sanitations in India with Special Reference to Cantonments*, Simla: Government Printing Office, 1901.

Rogers, Leonard, *On the Influence of Variations of the Ground Water Level on the Prevalence of Malarial Fevers,* Calcutta: no pub., 1897.

Scot, William, *Report of the Epidemic Cholera as It Has Appeared in the Territories Subject to the Presidency of Fort St. George,* Madras, 1824.

Snow, P.C.H., *Report on the Outbreak of Bubonic Plague in Bombay, 1896-97*, Bombay, 1897.

Stewart, Duncan, *Report on Smallpox in Calcutta, 1833-34, 1837-38, 1843-44 and Vaccination in Bengal from 1827 to 1844*, Calcutta: Government Press, 1844.

The Bengal Plague Manual, Being a Collection of the Extant and Executive Orders in Connection with Plague Issued by the Government of India and the Government of Bengal, Calcutta: Bengal Secretariat Press, 1903.

Townsend, S.C., *Memorandum on the Precautions to be Taken Against Cholera* (publication date unavailable).

Webster, W.J., *A Note on Hydrocyanic Acid Gas Fumigation of Rat Burrows as an Anti-Plague Measure*, New Delhi: Manager of Publications, 1935.

INDIA OFFICE LIBRARY, LONDON

1. Reports and Tracts

A Compilation of Regulations Issued by the Government of India and Local Governments in Connection with Plague. Calcutta: Home Department, 1898.

A Discourse on the Asiatic Cholera, and its Relations to Some Other Epidemics including General and Special Rules for its Prevention and Treatment. Thomas Henry Starr, London, 1848.

A Manual of Family Medicine and Hygiene for India, Sir William J. Moore, London, 1889.

Instructions in Regard to Cholera Epidemic. E.F. Barretto, Agra, 1892.

Medical Tracts: An Essay on the Cause, Diffusion, Location, Prevention, and Cure of the Asiatic Cholera and other Epidemics. William Schmoele, Philadelphia, 1866.

Memorandum on the Science and Art of Indian Medicine, Madras: Committee on the Indigenous Systems of Medicine, 1948.

Observations on Epidemics of Cholera in India, with Special Reference to their Immediate Connection with Pilgrimages. Charles Banks, Cuttack, 1896.

Report of an Enquiry into the Causes of Maternal Mortality in Calcutta. M.I. Neal Edwards, New Delhi: Government of India Publication 1940.v/25/850/94, Health Bulletin no. 27.

Report of Vaccination in Bengal during the Years 1893-96, H.J. Dyson, Calcutta, 1896. Appendix, 'Vaccination Among the Ferazi Mussalmans of Eastern Bengal'. A Note by Ben H. Deare.

Report on the Treatment of Epidemic Cholera, John Murray, Calcutta, 1869.

Simple Rules of Health for Young Officers Proceeding to India for the First Time. Major General J.B. Smith, India Office, 1926.

Statistics of Inoculations with Haffkin's Anti-Plague Vaccine, 1897-1900. W.B. Bannerman, Bombay, 1900

The Principal Diseases of India, Briefly Described; with Hints on the Duties of Medical Officers in that Country. Charles Alex Gordon, London, 1847.

The Bengal Plague Manual, Calcutta: Municipal Department, 1903.

2. Private Papers

Carey, Rev. William, 'Diary Including Records of His Experiences as a Missionary in India, 1870-1925'. MSS, Eur. B. 381, OIOC.

Fitzgerald, Patrick Gerald (1820-1910), *Diaries and Records of Service Mainly in Central India and Bengal, 1844-67*, 3 vols, A.140. OIOC.

Lamb, George (1869-1911), Major IMS and Director of the Pasteur Institute of India, 1905-09. Papers Related to Lamb, 1 Folder. MSS Eur D. 893, Oriental and India Office Collections (OIOC).

Letters, dated 1810-21 from Reginal Orton (d. 1835), telling his family about life in India and his career prospects, and about his book, *An Essay on the Epidemic Cholera of India*, MSS Eur. D 1036, OIOC.

Letter, dated 6 July 1866 from Florence Nightingale (1820-1910) mentioning the need for sanitary tracts written in India for India and her wish to obtain information on Indian educational matters, addressed to John Murdoch (1819-1904), writer and distributor of vernacular and English evangelical tracts and educational literature in Ceylon and India, MSS. Eur. A. 192, OIOC.

Short, H.E., 'In the Days of the Raj, and After: Doctor, Soldier, Scientist, Shikari', MSS Eur. C435, OIOC.

William, Mackie, 'The Control of Malaria in an Indian Cantonment', MSS Eur. C. 369, OIOC.

William Scott Collection. Henry Peers Dimmock (1857-1921), Lt. Col. Bombay, MS papers and two volumes of photographs entitled 'Plague Visitation Bombay 1896-97'. MSS Eur. D. 1148, OIOC.

3. Memoirs

Cleghorn, J., ed., *Scientific Memoirs by Medical Officers of the Army of India*, Calcutta, 1897. Indian Medical Research Memoir no. 32, Health Bulletin no. 28, *The Rice Problem in India*, Calcutta, 1933.

Indian Medical Research Memoirs, no. 34, March 1942. *A Comparative Nutritional Survey of Various Indian Communities,* D.C. Wilson and E.M. Widdowson. V/25/850/92.

Scientific Memoirs by Officers of the Medical and Sanitary Departments of the Government of India. *Investigation on Bengal Jail Dietaries with Some Observations on the Influence of Dietary on the Physical Development and Well Being of the People of Bengal.* Capt. D MacCay, Calcutta, 1910. V/25/850/50.

Rockefeller Archive Center, New York

Reports and Correspondences

Annual Report for the Year 1942, India, The Far East, by M.G. Belfour, RG 5.3, Series 464, Box 204, Folder 2485.

Bentley, C.A., Director of Public Health, to the General Director, IHB, Rockefeller Institution, New York, 7 November 1922, RF, IHB of the RF, Box 145, Folder, 2565, RG 5, Series 1.2, 1922.

————, Director of Public Health, to the General Director, IHB, RF, New York, 19 June 1924, Box 201, RG 5, Series 1.2, Bengal 466, Folder 2560.

Chatterjee, G.C., The Central Cooperative Anti-Malarial Society Ltd, through the Director of Public Health Government of Bengal (GOB) to the General Director International Health Board (IHB), Rockefeller Institute, New York, 28 August 1922, Rockefeller Foundation (RF), Box 14, Folder 2565, Record Group (RG) 5, Series 1.2 1922.

Dean, Rollin C., to J.F. Kendrick, 26 October 1925, RG 5, IHB/D, Series 1.2, Subseries 2, Box 232, Folder 2963.

Dean, Rollin, to K.S. Sitaram, 5 March 1936, RF, RG 2, 1936, Series 464, Box 138, Folder 1029.

Ferrell, John A., to Harry Timbres, RF, RG 2, 1932, Series 464, Box 74, Folder 598.

Heiser, Victor G., to C.A. Bentley, Box 201, RG 5, Series 1.2, Bengal 466, Folder 2560.

Heiser, Victor G., to J.F. Kendrick, Madras, India, IHB/D, Series 1.2, Subseries 2, Box 232, Folder 2962, 29 June 1925.

Kendrick, J.F., Madras to C.C. Williamson, New York, 24 November 1925, Series 1.2, Subseries 2, Box 232, Folder 2963.

Russell, F.F., Acting President to Sir Leonard Rogers, Honorable Medical Secretary, The British Empire Leprosy Relief Association, RF, Folder 2550-2559, RG 5, IBH/D, Series 1, Correspondence Subseries 2, Box 200, 464.

Sawyer, W.A., to C.A. Bentley, Director, Public Health for Bengal, 1925, RF Collection, RG 5, IHB/D, Series 1.2, Subseries 2, Box 232, Folder 2966.

Semi-Annual Report 1942, by Brian E. Dyer, All India Institute of Hygiene and Public Health, RF, RG 5, IHB/D, Series 3, Subseries Reports, Box 204, Folder 2484.

Sweet, W.C., RF, International Health Division (IHD) to W.A. Sawyer, RF,

New York, 9 May 1936, RG 2, 1936, Series Publications, Subseries 464, Box 138, Folder 1030

The British Empire Leprosy Relief Association to Dr. Wickliffe Rose, RF, Folder 2550-2559, RG 5, IBH/D, Series 1, Correspondence Subseries 2, Box 200, 464.

Watson, R.B., Notes on the Development of a Malaria Studies and Control Program by the Damodar Valley Corporation, RF, RG 2, 1950, Series General Correspondence, Subseries 464, Box 498, Folder 3336.

PERIODICALS

Aayurbignan Sammilani, Calcutta, 1931-2.
Bangabandhu, Dacca, 1875.
Bangajiban, Calcutta, 1896.
Chikitsak O Samalochak, Calcutta, 1895-1901.
Chikitsa-sammilani, Calcutta, 1887-94.
Krishaka, Calcutta, 1327-8 BS.
Natun Panjika, Calcutta, 1868.
Natun Panjika, Calcutta, 1871.
Natun Panjika, Serampur, 1868.
Pallibignan, Dacca, 1867-8.
Swasthya, Calcutta, 1898-1901.
Swasthya Samachar, Calcutta, 1912.
The Eastern Medical Bulletin, Calcutta, 1930.

B. SECONDARY SOURCES

Adamson, J., *The Cause and Cure of Asiatic or Malignant Cholera,* London: William and Thomas Piper, 1851.

Alavi, Seema, *Islam and Healing: Loss and Recovery of an Indo-Muslim Medical Tradition, 1600-1900*, Basingstoke: Palgrave MacMillan, 2008.

Annesley, J., *Sketches of the Most Prevent Diseases of India*, London: Underwood, 1825.

Arnold, David (ed.), *Imperial Medicine and Indigenous Societies,* Manchester: Manchester University Press, 1998.

————, 'Smallpox and Colonial Medicine in Nineteenth-Century British India', in David Arnold (ed.), *Imperial Medicine and Indigenous Societies,* Manchester: Manchester University Press, 1998.

————, 'Touching the Body: Perspectives on the Indian Plague, 1896-

1900', in *Subaltern Studies V*, ed. R. Guha, New Delhi: Oxford University Press, 1987, pp. 55-90.

———, 'An Ancient Race Outworn: Malaria and Race in Colonial India, 1860-1930', in W. Ernst and B. Harris (eds.), *Race, Science and Medicine, 1700-1960*, London: Routledge, 1999.

———, *Science, Technology and Medicine in Colonial India*, Cambridge: Cambridge University Press, 2000.

———, 'Cholera and Colonialism in British India', *Past and Present*, no. 113, 1986, pp. 118-51.

———, *Colonizing the Body: State Medicine and Epidemic Disease in Nineteenth-century India*, California: University of California Press, 1993.

———, *The Tropics and the Travelling Gaze: India, Landscape and Science, 1800-1856*, New Delhi: Permanent Black, 2005.

——— (ed.), *Warm Climates and Western Medicine: The Emergence of Tropical Medicine 1500-1900*, Amsterdam and Atlanta: Rodopi, 1996.

———, 'Disease, Resistance and India's Ecological Frontier, 1770-1947', Biswamoy Pati (ed.), *Issues in Modern Indian History*, Mumbai: Popular Prakashan, 2000, pp. 1-22.

Attewell, Guy, *Refiguring Unani Tibb: Plural Healing in Late Colonial India*, Hyderabad: Orient Longman, 2007.

Babb, Lawrence A., *The Divine Hierarchy: Popular Hinduism in Central India*, London: Columbia University Press, 1975.

Bala, Poonam (ed.), *Contesting Colonial Authority: Medicine and Indigenous Responses in Nineteenth- and Twentieth-century India*, New York, Toronto: Lexington Books, 2012.

——— (ed.), *Medicine and Colonialism: Historical Perspectives in India and South Africa*, London: Pickering and Chatto, 2014.

———, *Imperialism and Medicine in Bengal: A Socio-Historical Perspective*, New Delhi and London: Sage, 1991.

Bandopadhyay, Arun (ed.), *Science and Society in India, 1750-2000*, New Delhi: Manohar, 2010.

Bandyopadhyay, Bibhutibhushan, *Bibhuti Rachnabali*, Calcutta: Mitra Ghosh, 2000.

Bandyopadhyay, Haridhan, *Banglar Shatru*, Sodpur, publisher's name not available (hereafter no pub.), 1924.

Bandyopadhyay, Harinarayan, *Bala Chikitsa*, Calcutta: no pub., 1873.

Bandyopadhyay, Tarashankar, *Arogyoniketan*, Prakash Bhaban, Kolkata,

2007 (English translation by Enakshi Chatterjee, New Delhi: Sahitya Akademi, 1996).

Banerjee, Madhulika, *Power, Knowledge, Medicine: Ayurvedic Pharmaceuticals at Home and in the World*, Hyderabad: Orient Blackswan, 2009.

Banks, Charles, *Observations on Epidemics of Cholera in India, with Special Reference to their Immediate Connections with Pilgrimages*, Cuttack: no pub., 1896.

Baron, John, *The Life of Edward Jenner*, London: Henry Colburn, 1827.

Barretto, E.F., *Instructions in Regard to Cholera Epidemic*, Agra: no pub., 1892.

Basu, Amritakrishna, *Plague-tattva*, Calcutta, Taruni Press, 1899.

Basu, Gopendra Krishna, *Banglar Loukik Devata*, Calcutta: Ananda Publishers, 1964.

Bellew, H.W., *The History of Cholera in India from 1862 to 1881*, London: Trubner and Co., 1885.

Bewell, Allan, *Romanticism and Colonial Disease*, Baltimore: Hopkins University Press, 1999.

Bhaba, Homi K., *The Location of Culture*, London and New York: Routledge, 1994.

Bhardwaj, S.M., 'Homeopathy in India', in G.R. Gupta (ed.), *The Social and Cultural Context of Medicine in India*, New Delhi: Vikas Publishing House, 1981, pp. 1-54.

Bhattacharya, Sanjoy, 'Re-devising Jennerian Vaccines?: European Technologies, Indian Innovation and the Control of Smallpox in South Asia, 1850-1950', in Biswamoy Pati and Mark Harrison (eds.), *Health, Medicine and Empire: Perspectives on Colonial India*, New Delhi: Orient Longman, 2001.

———, *Expunging Variola: The Control and Eradication of Smallpox in India, 1947-77*, New Delhi: Orient Longman, 2006.

Bhattacharya, Sanjoy, Mark Harrison and Michael Worboys (eds.), *Fractured States: Smallpox, Public Health and Vaccination Policy in British India 1800-1947*, New Delhi: Orient Longman and Sangam Books, 2005.

Bhattacharyya, Ramnarayan Vidyaratna, *Gobij Prayog*, Calcutta: no pub., 1857.

Bose, Pradip Kumar (ed.), *Health and Society in Bengal: A Selection from Late 19th-century Bengali Periodicals*, New Delhi: Sage, 2006.

Bradfield, E.W.C., *An Indian Medical Review*, Delhi: Govt. of India Press, 1938.

Branca, Patricia, *The Medicine Show: Patients, Physicians and Perplexities of the Health Revolution in Modern Society,* New York: Science History Publications, 1977.

Briggs, Asa, 'Cholera and Society in the Nineteenth Century', *Past and Present,* no. 19, 1961, pp. 76-96.

Bynum, W.F., *Science and the Practice of Medicine in the Nineteenth Century,* Cambridge: Cambridge University Press, 1994.

Campbell, Oman John, *Cults, Customs and Superstitions of India,* Delhi: Vishal Publishers, 1972.

Catanach, I.J., 'Plague and the Tensions of Empire: India, 1896-1918', in David Arnold (ed.), *Imperial Medicine and Indigenous Societies,* Manchester: Manchester University Press, 1988, pp. 149-71.

————, 'Plague and the Indian Village, 1896-1914', in Peter Robb (ed.), *Rural India: Land, Power and Society Under British Rule,* London: Curzon Press, 1983, pp. 216-43.

Cave, A.J.E., 'Evidence of Incidence of Tuberculosis in Ancient Egypt', *British Journal of Tuberculosis,* vol. 33, 1935, pp. 142-52.

Chakrabarty, Dipesh, 'Shareer, Samaj O Rashtra: Oupanibeshik Bharatey Mahamari O Janasanskriti', in Gautam Bhadra and Partha Chattopadhyay, eds., *Nimnabarger Itihas,* Kolkata: Ananda Publishers, 2001.

Chakrabarti, Pratik, *Bacteriology in British India: Laboratory Medicine and the Tropics,* New York: Rochester University Press, 2012.

————, *Materials and Medicine: Trade, Conquest and Therapeutics in the Eighteenth Century,* Manchester: Manchester University Press, 2010.

————, *Western Science in Modern India: Metropolitan Methods, Colonial Practices,* New Delhi: Permanent Black, 2004.

————, 'Science, Nationalism, and Colonial Contestation: P.C. Ray and his Hindu Chemistry', *Indian Economic and Social History Review,* vol. 37, 2000, pp. 185-213.

Chakrabarti, Saradaprasad, *Visuchika-Chikitsasara,* Calcutta: Berini & Co., 1892.

Chandravarkar, Rajnarayan, 'Plague Panic and Epidemic Politics in India, 1896-1914', in T. Ranger and P. Slack (eds.), *Epidemics and Ideas: Essays on the Historical Perception of Pestilence,* Cambridge: Cambridge University Press, 1992, pp. 203-40.

Chattopadhyay, Saratchandra, *Sulabh Sarat Samagra,* Calcutta: Ananda Publishers, 1989.

Chevers, Norman, *A Commentary on the Diseases of India,* London: J. and A. Churchill, 1886.

Chopra, R.N., *Indigenous Drugs Inquiry: A Review of the Work*, Calcutta, Indian Research Fund Association, 1939.

Cook, Harold J., *Matters of Exchange: Commerce, Medicine, and Science in the Dutch Golden Age*, New Haven & London: Yale University Press, 2007.

Corbyn, F.A., *Treatise on the Epidemic Cholera as it has Prevailed in India*, Calcutta: Bengal Establishment, 1832.

Covell, G., 'Malaria in Calcutta', in *Records of Malaria Survey of India*, vol. 3, 1932, pp. 1-82.

Crawford, D.G., *A History of Indian Medical Service, 1600-1913*, London: London School of Tropical Medicine and Hygiene, 1914.

————, *Hughli Medical Gazetteer*, Calcutta: Bengal Secretariat Press, 1903.

Crosby, Alfred W., *Ecological Imperialism/The Biological Expansion of Europe, 900-1900*, Cambridge: Cambridge University Press, 1986.

Cuningham, J.M., *Cholera: What Can the State Do to Prevent it?* Calcutta: Supt. of Govt. Printing, 1884.

Cunningham, A. and B. Andrews (eds.), *Western Medicine as Contested Knowledge*, Manchester: Manchester University Press, 1997.

Curtin, Philip D., *Death by Migration: Europe's Encounter with the Tropical World in the Nineteenth Century*, Cambridge: Cambridge University Press, 1989.

————, 'The End of the "White Man's Grave"? Nineteenth-century Mortality in West Africa', *Journal of Interdisciplinary History*, vol. 21, no. l, 1990, pp. 63-88.

Das, Charuchandra, *Banglar Samasya*, Calcutta: no pub., 1932.

Das, Nilmoni, *Olauthar Chi kits a Pranali*, Calcutta: no pub., 1877.

Dasgupta, Nagendrachandra, *Banglar Palii Samasya*, Calcutta: no pub., 1932.

Datta, Saratchandra, *Olautha-Chikitsa*, Calcutta: no pub., 1877.

Dawtrey, Drewitt, *The Life of Edward Jenner*, London: Longmans Green and Co., 1931.

De, S.N., *Cholera: Its Pathology and Pathogenesis*, Edinburgh and London: Oliver and Boyd, 1961.

————, 'Anopheline Survey and Malaria Fever in Calcutta', in *Indian Medical Records*, vol. 43, 1923, pp. 86-9.

Deb Roy, Rohan, '"An Unseen Awful Visitant": The Making of Burdwan Fever', *Economic and Political Weekly*, vol. 43, nos. 12 and 13, 22 March-4 April 2008.

Delaporte, F.J., *Disease and Civilization: The Cholera in Paris, 1832*, Cambridge: MIT Press, 1986.

Desai, U.L., *Plague in India,* Simla, Government Central Printing Office, 1898.

Dey, Rev. Lal Behari, *Bengal Peasant Life,* Calcutta: Firma K.L.M. 1969.

Dey, Kanny Lall, *Hindu Social Laws and Habits Viewed in Relation to Health,* Calcutta: Thacker, Spink & Co., 1866.

———, *Indigenous Drugs in India,* Calcutta: Thacker, Spink & Co., 1867.

Dharampal, *Indian Science and Technology in the Eighteenth Century: Some Contemporary European Accounts*, Delhi: Impex India, 1971.

Dhur, Gobin Chunder, *The Plague, Being a Reprint of Letters Published in the Indian Mirror for Allaying Popular Alarm and, Conciliating the People to the Action of the Authorities*, Calcutta: Sanyal & Co., 1898.

Douglas, Mary, *Essays in the Sociology of Perception,* London: Routledge and Kegan Paul, 1982.

Dubey, S.C., *Indian Village,* London: Routledge and Kegan Paul, 1955.

Durey, M., *The Return of the Plague: British Society and the Cholera, 1831-32*, Dublin: Gill and Macmillan, 1979.

Dutta, Abhijit et al. (eds.), *Explorations in History: Essays in Honour of Professor C. Palit,* Kolkata: Corpus Research Institute, 2003.

Edmondson, Charles T., *Popular Information on Smallpox, Inoculation and Vaccination*, Calcutta: no. pub., 1870.

Ernst, W. and B. Harris (eds.), *Race, Science and Medicine, 1700-1960,* London: Routledge, 1999.

Ernst, Waltraud (ed.), *Plural Medicine, Tradition and Modernity, 1800-2000*, London: Routledge, 2002.

Evans, Richard J., 'Epidemics and Revolutions: Cholera in Nineteenth-Century Europe', *Past and Present,* vol. 1, no. 20, 1988, pp. 123-46.

Foucault, Michel, *The Birth of the Clinic: An Archaeology of Medical Perception*, London: Routledge, 2003.

Gait, E.A., *A History of Assam,* Calcutta: Thacker, Spink & Co., 1906.

Gallagher, Nancy, *Medicine and Power in Tunisia, 1780-1900,* New York: Cambridge: Cambridge University Press, 1983.

Garrison, F.H., *An Introduction to the History of Medicine,* Philadelphia: W.B. Saunders and Co. 1929.

Ghose, Benoy Krishna, *A Brief Miscellany of Indian Domestic Medicines,* Calcutta: no pub., 1887.

Ghosh, Dayal Krishna, *Malaria: A Treatise on its Origin and Treatment,* Calcutta: no pub., 1878.

Ghosh, Radhakanta, *Homaeopathic Mate Rog Chikitsa: Olautha,* Calcutta: no pub., 1877.

Ghosh, Santosh Kumar, *Jaksharog o Pratirodh,* Calcutta, no pub., 1363 (BS).

Gordon, C.A., *Notes on the Hygiene of Cholera,* Madras, Balliere, Tindall, and Co., 1877.

———, *The Principal Diseases of India Briefly Described,* London: no pub., 1847.

Greenough, Paul, 'Intimidation, Coercion and Resistance in the Final Stages of the South Asian Smallpox Eradication Campaign, 1973-75', *Social Science and Medicine,* vol. 4, no. 5, 1995, pp. 663-85.

Griffith, Ralph T.H., *The Hymns of the Atharva Veda,* Varanasi: Chowkhamba Sanskrit Series Office, 1968.

Guha, Ranajit and Gayatri Chakraborty Spivak (eds.), *Selected Subaltern Studies,* Oxford: Oxford University Press, 1988.

Guha, Sumit, *Health and Population in South Asia: From Earliest Times to the Present,* London: C. Hurst, 2001.

Gupta, B., 'Indigenous Medicine in Nineteenth and Twentieth-century Bengal', in *Asian Medical Systems: A Comparative Study,* Delhi: Motilal Banarsidass, 1998, pp. 368-78.

Hankin, E.H., *The Cause and Prevention of Cholera,* Agra: Star Press, 1897.

Hardiman, David, *The Coming of the Devi: Adivasi Assertion in Western India,* New Delhi: Oxford University Press (New Edition), 1995.

Hardiman, David and Projit Bihari Mukharji, 'Subaltern Therapeutics in Indian Medical History', *Wellcome History, 9* July 2014.

Harrington, B.R., *Sanitary Engineering for India,* Calcutta: Thacker, Spink & Co., 1887.

Harrison, Mark, *Climates and Constitutions: Health, Race, Environment and British Imperialism in South Asia, 1600-1850,* New Delhi, Oxford University Press, 1999.

———, *Public Health in British India: Anglo-Indian Preventive Medicine 1859-1914,* Cambridge: Cambridge University Press, 1994.

———, 'A Question of Locality: The Identification of Cholera in British India, 1860-1890', in David Arnold (ed.), *Warm Climates and Western Medicine,* Amsterdam: Rodopi, 1996, pp. 133-59.

———, 'Tropical Medicine in Nineteenth-Century India', *The British Journal for the History of Science,* vol. 25, 1992, pp. 299-318.

Harrison, Mark and Michael Worboys, 'A Disease of Civilization: Tuberculosis in Britain, Africa and India, 1900-1939', in M. Worboys and L. Marks (eds.), *Migrants, Minorities and Health: Historical and Contemporary Studies,* London: Routledge, 1997.

Hart, Ernest, *The Medical Profession in India: Its Position and Its Work.* An Address Delivered Before the Indian Medical Congress held at Calcutta in December 1894, Calcutta: Thacker, Spink & Co., 1895.

Hays, J.N., *Epidemics and Pandemics: Their Impacts on Human History*, Santa Barbara, California: ABC-CLIO, 2005.

Headrick, Daniel R., *Tools of Empire: Technology and European Imperialism in the Nineteenth Century*, Oxford: Oxford University Press, 1981.

————, *The Tentacles of Progress: Technology Transfer in the Age of Imperialism, 1850-1940*, New Oxford: Oxford University Press, 1988.

Hewitt, F., *The History of British Settlements in India*, London: no pub., 1855.

Holwell, J.Z., *An Account of the Manner of Inoculating for the Smallpox in the East Indies*, London: T. Becket and P.A. De Hondt, 1767.

Iyengar, T.C. Rajan, *A Brief Sketch of the Symptoms and Treatment of Cholera*, Poona: no pub., 1892.

James, S.P., *Smallpox and Vaccination in British India*, Calcutta: Thacker, Spink & Co., 1909.

Jeffrey, Roger, *The Politics of Health in India*, Berkeley, Los Angeles and London: University of California, 1988.

Jones, W.H.S., *Malaria: A Neglected Factor in the History of Greece and Rome*, London: School of Tropical Medicine and Hygiene, 1907.

Kakar, D.N., *Primary Health Care and Traditional Medical Practitioners*, New Delhi: Munshiram Manoharlal, 1988.

Kakar, Sudhir, *Shamans, Mystics and Doctors: A Psychological Enquiry into India and its Healing Traditions*, New York: A. Knopf, 1982.

Kamat, Manjiri, '"The Palkhi as Plague Carrier": The Pandharpur Fair and the Sanitary Fixation of the Colonial State: British India, 1908-16', in Biswamoy Pati and Mark Harrison (eds.), *Health, Medicine and Empire: Perspectives on Colonial India*, New Delhi: Orient Longman, 2001, pp. 299-316.

Kaviraja, Haradhan Vidyaratna, *Basanta Roger Nidan O Chikitsa*, Calcutta: no pub., 1868.

Kazi, Ihtesham, *A Historical Study of Malaria in Bengal, 1860-1920*, Dhaka: Pip International Publications, 2004.

Kennedy, R.H., *Notes on the Epidemic Cholera*, Calcutta: Baptist Mission Press, 1827.

Klein, E.R. and H. Gibbes, *Inquiry into the Etiology of Asiatic Cholera*, 1885, no place of publication and no pub.

Klein, Ira, 'Plague, Policy and Popular Unrest in British India', *Modern Asian Studies*, vol. 22, no. 4, 1988, pp. 728-53.

————, 'Malaria and Mortality in Bengal, 1840-1921', *Indian Economic and Social History Review*, vol. 9, no. 2, 1972, pp. 132-60.

————, 'Death in India, 1871-1921', *Journal of Asian Studies,* vol. 32, no. 4, pp. 639-59.

————, 'Imperialism, Ecology and Disease: Cholera in India, 1850-1950', *The Indian Economic and Social History Review,* vol. 31, no. 4, 1994, pp. 491-518.

Kumar, Anil, *Medicine and the Raj: British Medical Policy 1835-1911,* New Delhi: Sage, 1998.

Kumar, Deepak (ed.), *Disease and Medicine in India: A Historical Overview,* New Delhi: Tulika, 2001.

————, *Science and the Raj, 1857-1905,* New Delhi: Oxford University Press, 1995.

———— (ed.), *Science and Empire: Essays in Indian Context,* New Delhi: Anamika Prakashan, 1991.

————, 'Social History of Medicine; Some Issues and Concerns', in Deepak Kumar (ed.), *Disease and Medicine in India: A Historical Overview,* New Delhi: Tulika, 2001, pp. xiv-xvii.

Kumar, Deepak and Raj Sekhar Basu (eds.), *Medical Encounters in British India,* New Delhi: Oxford University Press, 2013.

Lamb, G., *The Etiology and Epidemiology of Plague: A Summary of the Work of the Plague Commission*; issued under the Authority of the Government of India by the Sanitary Commissioner with the Government of India, Calcutta: Superintendent Government Printing, 1908.

Leslie, C. (ed.), *Asian Medical Systems: A Comparative Study,* Delhi, Motilal Banarsidass, 1998.

MacLean, W.C., *Diseases of Tropical Climates,* London: Macmillan, 1886.

Macgregor, W.L., *Practical Observations on the Principal Diseases Affecting European and Native Soldiers in the NW Provinces of India*, Calcutta: W. Thacker, 1843.

Macleod, Roy and Milton Lewis (eds.), *Disease, Medicine and Empire,* London: Routledge, 1988.

Macnamara, C., *A History of Asiatic Cholera,* London: Macmillan, 1876.

————, *A Treatise on Asiatic Cholera*, London: Churchill, 1870.

Macpherson John, K., *Annals of Cholera: From the Earliest Periods to the Year 1817,* London: Ranken and Drury, 1872.

Majumdar, Gopal Chandra, *Tikadarganer Prati Upadesh,* Calcutta: no pub., 1872.

Mandal, Rajkrishna, *Malariar Karan O Pratikar,* Calcutta, 1908.

Marshman, John Clark, *The History of India from the Earliest Period to the Close of Lord Dalhousie's Administration*, London: Longmans, Green, Reader and Dyer, 1867.

May, J.M., *The Ecology of Human Disease,* New York: M.D. Publications, 1958.

McGrew, R.E., *Russia and the Cholera, 1823-32,* Madison: University of Wisconsin Press, 1965.

Menon, P. Karunakaran, *An Essay on Plague,* Coimbatore: no pub., 1898.

Metcalf, T., *Ideologies of the Raj,* Cambridge: Cambridge University Press, 1994.

Meulenbeld, G.J. and Dominik Wujastyk (eds.), *Studies on Indian Medical History,* Delhi: Motilal Banarsidass, 2001.

Minsky, Lauren, 'Pursuing Protection from Disease: The Making of Smallpox Prophylactic Practice in Colonial Punjab', *Bulletin of the History of Medicine,* vol. 83, no. 1, 2009, pp. 164-90.

Misra, Bhabagrahi, 'Sitala: The Smallpox Goddess of India, *Asian Folklore Studies,* vol. 28, no. 2, 1969, pp. 133-42.

Mitra, A., *Cholera and its Cure,* Lahore: no pub., 1888.

———, *The Bubonic Plague,* Calcutta, Thacker, Spink & Co., 1897.

Mitra, Girindra Krishna, *Skeleton of a Scheme to Combat Malaria and other Prevalent Diseases in Bengal,* Calcutta: Department of Public Health, 1925.

———, *Malaria O Kalajwarer Pritikar Samasyer Parikalpana,* Calcutta: no pub., 1924.

Moore, James C., *The History of the Smallpox,* London: School of Hygiene and Tropical Medicine, 1815.

Mukharji, P.B., *Nationalizing the Body: The Medical Market, Print and Daktari Medicine,* London: Anthem Press, 2009.

———, 'In-Disciplining Jwarasur: The Folk/Classical and Trans-materiality of Fevers in Colonial Bengal', *Indian Economic and Social History Review,* vol. 50, no. 3, 2013, pp. 261-88.

———, 'The "Cholera Cloud" in the Nineteenth Century British World: History of an Object-without-an Essence', *Bulletin for the History of Medicine,* vol. 86, no. 3, 2012, pp. 303-32.

Mukhaty, H.N., *Cholera Guide,* Dacca: no pub., 1924.

Mukherjee, Nilmoni, *A Bengal Zamindar: Jaykrishna Mukherjee of Uttarpara and His Times, 1809-88,* Calcutta: K.P. Bagchi and Co., 1975.

Mukhopadhyay, Arun Kumar, *Cholera Chikitsa,* Calcutta: no pub., 1330 (BS).

Mukhopadhyaya, Paresnath, *Olautha,* Calcutta: no pub., 1873.

Mukhopadhyaya, Jadunath, *Kuinayn Prayagapranali,* Chinsura: Chikitsa Prakash Press, 1873.

———, *Bisuchika Roger Chikitsa,* Chinsura: Chikitsa Prakash Press, 1872.

Muraleedharan, V.R., 'Malady in Madras: The Colonial Government's

Responses to Malaria in the Early Twentieth Century', in Deepak Kumar (ed.), *Science and Empire: Essays in Indian Context,* New Delhi: Anamika, 1990, pp. 101-14.

Murdoch, J., *Sanitary Reforms in India,* Madras: Christian Literature Society, 1888.

Nag, Jnanendra Krishna, *Cholera Chikitsa,* Calcutta: no pub., 1917.

Naraindas, Harish, 'Care, Welfare, and Treason: The Advent of Vaccination in the 19th Century', *Contributions to Indian Sociology,* vol. 32, no. 1, 1998, pp. 67-96.

Nathan, R., *The Plague in India,* Simla, Government Central Printing Office, 1898.

Nicholas, Ralph W., 'The Goddess Sitala and Epidemic Smallpox in Bengal', *Journal of Asian Studies,* vol. 41, no. 1, 1981, pp. 21-41.

Nostrand, D. Van, *Social Systems: Their Persistence and Change,* Princeton: D. Van Nostrand Co., 1960.

Oldham, C.F., *What is Malaria and Why is it Most Intense in Hot Climates?* London: H.K. Lewis, 1871.

Pahari, Subrata, *Unish Shataker Banglay Sanatani Chikitsa Byabasthar Swarup,* Kolkata: Progressive Publishers, 2001.

Pati, Biswamoy and Mark Harrison (eds.), *The Social History of Health and Medicine in Colonial India,* London and New York: Routledge, 2009.
————, *Health, Medicine and Empire: Perspectives on Colonial India,* New Delhi: Orient Longman, 2001.

Paul, B.D. (ed.), *Health Culture and Community,* New York: Russell Sage Foundation, 1951.

Peers, Douglas M., 'Soldiers, Surgeons and the Campaigns to Combat Sexually Transmitted Diseases in Colonial India, 1805-1860', *Medical History,* no. 42(2), April 1998, pp. 137-60.

Pointer, E.N.L. (ed.), *Medicine and Culture,* London: no pub., 1969.

Power, Helen J., 'The Calcutta School of Tropical Medicine: Institutionalizing Medical Research in the Periphery', *Medical History,* vol. 40, 1996, pp. 197-214.

Prakash, Gyan, *Another Reason: Science and the Imagination of Modern India,* Princeton: Princeton University Press, 1999.

Quaiser, Neshat, 'Politics, Culture and Colonialism: Unani's Debate with Doctory', in Biswamoy Pati and Mark Harrison (eds.), *Health Medicine and Empire: Perspectives on Colonial India,* Hyderabad: Orient Longman, 2001, pp. 317-55.

Ramanna, Mridula, *Western Medicine and Public Health in Colonial Bombay, 1845-95,* Hyderabad: Orient Longman, 2002.

Ramasubban, Radhika, *Public Health and Medical Research in India: Their Origins and Development under the Impact of British Colonial Policy*, Stockholm: SAREC, 1982.

Ray, Kabita, *History of Public Health: Colonial Bengal 1921-1947*, Calcutta: K.P. Bagchi and Co., 1998.

————, 'Smallpox and the Introduction of Vaccination in Colonial Bengal', in Abhijit Dutta et al. (eds.), *Explorations in History: Essays in the Honour of Professor Chittabrata Palit*, Kolkata: Corpus Research Institute, 2003, pp. 106-16.

Ray, Karalicharan, *Bange Malaria*, Vasantapur: no pub., 1917.

Read, Margaret, *Culture, Health and Disease: Social and Cultural Indifferences on Health Programmes in Developing Countries*, London: Tavistock Publications, 1966.

Rogers, Sir Leonard, 'Gleanings from the Calcutta Post-Mortem Records: II. The Incidence of Tuberculosis Disease in Bengal', *Indian Medical Gazette*, February 1909.

————, *Smallpox and Climate in India: Forecasting of Epidemics*, London: HMSO, 1926.

Rogers, Sir Leonard, *Fevers in the Tropics*, London: Henry Frowde, 1919.

Rosenberg, Charles E. and Janet Golden (eds.), *Framing Disease: Studies in Cultural History*, New Brunswick: Rutgers University Press, 1992.

————, *The Cholera Years: The United States in 1832, 1849 and 1866*, Chicago: University of Chicago Press, 1962.

Ross, Ronald, *The Prevention of Malaria*, London: J. Murray, 1910.

————, *Memoirs*, London: J. Murray, 1923.

Roy, Binaybhushan, *Unish Shatake Banglar Chikitsa Byabastha, Deshiya Bheshaj O Sarkar*, Delhi: International Centre for Bengali Studies, 1998.

Roy, G.C., *The Causes, Symptoms and Treatment of Burdwan Fever*, Calcutta: Thacker, Spink & Co., 1876.

Salzer, Leopold, *Lectures on Cholera*, Calcutta: C. Ringer & Co., 1937.

Samanta, Arabinda, *Malarial Fever in Colonial Bengal, 1820-1947: Social History of an Epidemic*, Calcutta: Firma KLM, 2002.

————, 'Plague and Prophylactics: Ecological Construction of an Epidemic in Colonial India', in Ranjan Chakrabarti (ed.), *Situating Environmental History*, New Delhi: Manohar, 2007, pp. 221-44.

————, 'Plague and Prophylactics: Interrogating Colonial Medical Intervention in Eastern India', in Arun Bandopadhyay (ed.), *Science and Society in India, 1750-2000*, New Delhi: Manohar, 2010, pp. 121-44.

————, 'Smallpox in Nineteenth Century Bengal', *Indian Journal of History of Science,* vol. 47, no. 2, 2012, pp. 211-40.

————, 'Malarial Fever in Nineteenth-Century Bengal: Revisiting the Prophylactic Intervention', in Poonam Bala (ed.), *Contesting Colonial Authority: Medicine and Indigenous Responses in Nineteenth- and Twentieth-century India,* Lanham, Boulder, New York, Toronto, Plymouth: Lexington Books, 2012, pp. 137-52.

————, 'Re-visiting an Indigenous Medical Practice: The *Tikadars* in Colonial Bengal', in Syed Ejaz Hussain and Mohit Saha (eds.), *India's Indigenous Medical Systems: A Cross-Disciplinary Approach,* Delhi: Primus Books, 2015, pp. 151-62.

Sarkar, Nalini K., *Malaria: Its Causation and Means of Preventing it,* Calcutta: no pub., 1917.

Sarkar, Mahendra Lal, *A Sketch of the Treatment of Cholera,* Calcutta: P. Sarkar Anglo-Sanskrit Press (2nd edn.), 1904.

Sarkar, Simkie, 'Malaria in Nineteenth-century Bombay', in Deepak Kumar (ed.), *Disease and Medicine in India: A Historical Review,* New Delhi: Tulika, 2001, pp. 132-43.

Scheube, B., *The Diseases of Warm Countries: A Manual for Medical Men,* London: John Bale, Sons & Danielsson, Ltd., 1903.

Sen, Indubhusan, *Bangalir Khadya,* Calcutta: no pub., 1932.

Shatri, Shibnath, *Ramtanu Lahiri O Tatkalin Banga Samaj,* ed. Baridbaran Ghosh, Kolkata: New Age Publications, 2007.

Shattuck, G.C., *Diseases of the Tropics,* New York: Appleton-Century-Crofts, 1951.

Simpson, J.Y., *Proposal to Stamp out Smallpox and Other Contagious Diseases,* Edinburgh: Edmonston and Douglas, 1868.

Singh, Dhrubkumar, '"Clouds of Cholera" and Clouds around Cholera, 1817-70', in Deepak Kumar (ed.), *Disease and Medicine in India: A Historical Overview,* New Delhi: Tulika, 2001, pp. 144-65.

Sirkar, Mahendra Lal, *Therapeutics of Plague, Being Suggestions for the Prophylactic and Curative Treatment of the Disease,* Calcutta: no pub., 1899.

Sleeman, W.H., *Rambles and Recollections of an Indian Official,* vol. 1, London: J. Hatchard & Son, 1844.

Sorby, Arnold, *Medicine and Mankind,* London: Faber and Faber, 1941.

Starr, Thomas Henry, *A Discourse on the Asiatic Cholera, and its Relations to Some Other Epidemics including General and Special Rules for its Prevention and Treatment,* London: John Churchill, 1848.

Stewart, Tony, *Encountering the Smallpox Goddess: The Auspicious Song of Sitala,* Princeton: Princeton University Press, 1995.

Swanson, M.W., 'The Sanitation Syndrome: Bubonic Plague and Urban Native Policy in the Cape Colony, 1900-1909', *Journal of African History*, no. 18, 1977, pp. 387-410.

Tagore, Rabindranath, *Rabindra Rachnabali*, Calcutta: Visva Bharati, 1398 BS.

Thomas, Keith, *Religion and the Decline of Magic: Studies in Popular Beliefs in Sixteenth- and Seventeenth-century England*, London: Penguin Books, 1980.

Ukil, A.C., 'Anti-Tuberculosis Work in Bengal', *Indian Medical Gazette*, vol. 73, 1938, pp. 525-9.

Vidyaratna, Haradhan, *Basanta Roger Nidana*, Calcutta: no pub., 1848, 1868.

Vyas, B.J., *Notes on Sanitary Primer*, Ahmedabad: no. pub., 1886.

Wadley, Susan, 'Sitala, the Cool One', *Asian Folklore Studies*, vol. 39, no. 1, pp. 33-62.

Wakimura, Kohei, 'Malaria Control under the Colonial Rule: India and Taiwan', *South Asia: Institutions, Changes and Networks Quarterly Journal*, vol. 1, no. 3, 1999, pp. 9-25.

Wall, A.J., *Asiatic Cholera*, London: H.K. Lewis, 1893.

Watts, Sheldon J., 'British Development Policies and Malaria in India, 1897-c. 1929', *Past and Present*, vol. 165, 1999, pp. 141-81.

White, J., *Treatise on Cholera Morbus; The Method of Treatment and Means of Prevention*, London: W. Strange, 1834.

Wilson, D.C. and E.M. Widdowson, *A Comparative Survey of Various Indian Communities*, Calcutta: Thacker, Spink & Co., 1942.

Wildavsky, Aaron and Karl Drake, 'Theories of Risk Perception: Who Fears What and Why?', in Susan L. Cutter (ed.), *Environmental Risks and Hazards*, New Jersey: Prentice Hall, 1994, pp. 166-77.

Wilson, R.N., *The Sociology of Health: An Introduction*, New York: Random House, 1970.

Winslow, C.E.A., *Man and Epidemics*, Princeton: Princeton University Press, 1952.

Wise, T.A., *Commentary on the Hindu System of Medicine* (2nd edn.), London: Trubner and Co., 1860.

Worboys, Michael, 'The Emergence of Tropical Medicine: A Study in the Establishment of a Scientific Specialty', in Gerard Lemaine, Roy MacLeod, Michael Mulkay and Peter Weingart (eds.), *Perspectives on the Emergence of Scientific Disciplines*, Aldine: Chicago, 1976, pp. 75-98.

————, 'The Sanatorium Treatment for Consumption in Britain, 1890-1914', in John V. Pickstone (ed.), *Medical Innovations in Historical Perspective,* London: Macmillan, 1992, pp. 47-71.

Wujastyk, Dominik, 'A Pious Fraud': The Indian Claims for Pre-Jennerian Smallpox Vaccination', in G.J. Meulenbeld and Dominik Wujastyk (eds.), *Studies on Indian Medical History*, Delhi: Motilal Banarsidass, 2001.

Index

Printed and bound by CPI Group (UK) Ltd, Croydon, CR0 4YY

17/10/2024

01775682-0008